Philosophy of Love and Sex—Third Edition

Philosophy of Love and Sex—Third Edition

edited by Suzanne Senay

Canadian Scholars' Press Inc.
Custom Textbook Series
Toronto

Philosophy of Love and Sex—Third Edition © 2014 by Suzanne Senay
First published in 2010 by:
Canadian Scholars' Press Inc.
425 Adelaide Street West, Suite 200
Toronto, Ontario M5V 3C1
1.800.463.1998
www.coursepack.ca

LIBRARY AND ARCHIVES CANADA CATALOGUING IN PUBLICATION

Senay, Suzanne.
Philosophy of Love and Sex.

ISBN 978-1-55130-563-9
1. Philosophy — Sexuality — Introduction. I. Title.

Cover photo: "Psyche Revived by Cupid's Kiss"; Hermitage Museum, St. Petersburgh © 1787, Antonio Canova.

Cover design and interior layout: Benjamin Craft

Printed and bound in Canada by Webcom.

Table of Contents

What Is Philosophy?

1 *Republic* Book VII: The Myth of the Cave — 1
Plato

Concepts of Love

2 Six Types of Love — 5
Eric Fromm

3 "Varieties of Love: What is Love?" — 11
Alan Soble

Self-Love

4 *Nichomachean Ethics:* Self Love — 31
Aristotle

Erotic Love and Sexual Desire

5 *Theogony* — 35
Hesoid

6 Fragments: 31, 94, 130, 147 — 45
Sappho

7 *Symposium:* Speech of Aristophanes — 47
Plato

8 *Symposium:* Speech of Socrates — 53
Plato

9 *The Hebrew Bible:* Song of Solomon — 63

10 *Lyrics Rude & Erotic*: Poems 5, 7, 32, 75 73
Catullus

11 *Confessions*: Book II, 1-10; Book III, 1 75
St. Augustine

12 *The Holy Qur'an*: Sura 4: Women 83

13 "Love Supreme: Gay Nuptials and the Making of Modern Marriage" 89
Adam Haslett

14 *Lust:* "Hobbesian Unity" and "Disasters" 97
Simon Blackburn

15 Sonnet XXIX, CXIV 105
William Shakespeare

16 "For Anne Gregory" 107
W.B. Yeats

17 "Sex Without Love" 109
Sharon Olds

Friendship

18 *Nichomachean Ethics:* Friendship 111
Aristotle

19 *The Gay Science:* §14, §192, §334 119
Friedrich Nietzsche

Family Love

20 *Analects:* Book I 121
Confucius

21 *The Reasons of Love* 125
Harry Frankfurt

Divine Love and Altruism

22 *The New Testament:* First Letter of Paul to the Corinthians 135

23 *The Teachings of the Compassionate Buddha:* The Bodhisattva's Vow of Universal Redemption 137
E.A. Burtt (Ed.)

24 *Confessions*: Book III, 3-6; Book X, 6-7, 26-29 141
St. Augustine

25 *The Life of St. Teresa by Herself* 147
St. Teresa

26 *Civilization and its Discontents* 157
Sigmund Freud

Sex, Nature, and Normativity

27 *The Hebrew Bible:* Genesis 165

28 *The Hebrew Bible:* Leviticus 173

29 *Summa Theologica* 175
Thomas Aquinas

30 *Philosophy in the Bedroom* 183
Marquis de Sade

31 "Sexual Perversion" 191
Thomas Nagel

32 "The Human Sexual Response Cycle" 201
Sinclaire Intimacy Institute

33 Sex is Not a Natural Act: "Historical, Scientific, Clinical, and Feminist Criticisms of the 'Human Sexual Response Cycle' Model" 205
Leonore Tiefer

34 Sex is Not a Natural Act: "The Kiss" 221
Lenore Tiefer

Homosexuality

35 *The New Testament*: Letter of Paul to the Romans 227

36 *The Holy Qur'an*: Sura 27: "The Ants", Sura 29: "The Spider" 231

37 *The Morality of Homosexuality* 239
Michael Ruse

Acknowledgements 247

1 *Republic*, Book VII: The Myth of the Cave

Plato

Note: The first speaker is Socrates. The respondent is Glaucon.

Next, I said, compare the effect of education and the lack of it upon our human nature to a situation like this: imagine men to be living in an underground cave-like dwelling place, which has a way up to the light along its whole width, but the entrance is a long way up. The men have been there from childhood, with their neck and legs in fetters, so that they remain in the same place and can only see ahead of them, as their bonds prevent them turning their heads. Light is provided by a fire burning some way behind and above them. Between the fire and the prisoners, some way behind them and on a higher ground, there is a path across the cave and along this a low wall has been built, like the screen at a puppet show in front of the performers who show their puppets above it. — I see it.

See then also men carrying along that wall, so that they overtop it, all kinds of artifacts, statues of men, reproductions of other animals in stone or wood fashioned in all sorts of ways, and, as is likely, some of the carriers are talking while others are silent. — This is a strange picture, and strange prisoners.

They are like us, I said. Do you think, in the first place, that such men could see anything of themselves and each other[1] except the shadows which the fire casts upon the wall of the cave in front of them? — How could they, if they have to keep their heads still throughout life?

And is not the same true of the objects carried along the wall? — Quite.

If they could converse with one another, do you not think that they would consider these shadows to be the real things? — Necessarily.

What if their prison had an echo which reached them from in front of them? Whenever one of the carriers passing behind the wall spoke, would they not think that it was the shadow passing in front of them which was talking? Do you agree? — By Zeus I do.

Altogether then, I said, such men would believe the truth to be nothing else than the shadows of the artifacts? — They must believe that.

Consider then what deliverance from their bonds and the curing of their ignorance would be if something like this naturally happened to them. Whenever one of them was freed, had to stand up suddenly, turn his head, walk, and look up toward the light, doing all that would give him pain, the flash of the fire would make it impossible for him to see the objects of which he had earlier seen the shadows. What do you think he would say if he was told that what he saw then was foolishness, that he was now somewhat closer to reality and turned to things that existed more fully, that he saw more correctly? If one then pointed to each of the objects passing by, asked him what each was, and forced him to answer, do you not think he would be at a loss and believe that the things which he saw earlier were truer than the things now pointed out to him? — Much truer.

If one then compelled him to look at the fire itself, his eyes would hurt, he would turn round and flee toward those things which he could see, and think that they were in fact clearer than those now shown to him. — Quite so.

And if one were to drag him thence by force up the rough and steep path, and did not let him go before he was dragged into the sunlight, would he not be in physical pain and angry as he was dragged along? When he came into the light, with the sunlight filling his eyes, he would not be able to see a single one of the things which are now said to be true. — Not at once, certainly.

I think he would need time to get adjusted before he could see things in the world above; at first he would see shadows most easily, then reflections of men and other things in water, then the things themselves. After this he would see objects in the sky and the sky itself more easily at night, the light of the stars and the moon more easily than the sun and the light of the sun during the day. — Of course.

Then, at last, he would be able to see the sun, not images of it in water or in some alien place, but the sun itself in its own place, and be able to contemplate it. — That must be so.

After this he would reflect that it is the sun which provides the seasons and the years, which governs everything in the visible world, and is also in some way the cause of those other things which he used to see. — Clearly that would be the next stage.

What then? As he reminds himself of his first dwelling place, of the wisdom there and of his fellow prisoners, would he not reckon himself happy for the change, and pity them? — Surely.

And if the men below had praise and honours from each other, and prizes for the man who saw most clearly the shadows that passed before them, and who could best remember which usually came earlier and which later, and which came together and thus could most ably prophesy the future, do you think our man would desire those rewards and envy those who were honoured and held power among the prisoners, or would he feel, as Homer put it, that he certainly wished to be "serf to another man without possessions upon the earth"[2] and go through any suffering, rather than share their opinions and live as they do? — Quite so, he said, I think he would rather suffer anything.

Reflect on this too, I said. If this man went down into the cave again and sat down in the same seat, would his eyes not be filled with darkness, coming suddenly out of the sunlight? — They certainly would.

And if he had to contend again with those who had remained prisoners in recognizing those shadows while his sight was affected and his eyes had not settled down — and the time for this adjustment would not be short — would he not be ridiculed? Would it not be said that he had returned from his upward journey with his eyesight spoiled, and that it was not worthwhile even to attempt to travel upward? As for the man who tried to free them and lead them upward, if they could somehow lay their hands on him and kill him, they would do so. — They certainly would.

This whole image, my dear Glaucon, I said, must be related to what we said before. The realm of the visible should be compared to the prison dwelling, and the fire inside it to the power of the sun. If you interpret the upward journey and the contemplation of things above as the upward journey of the soul to the intelligible realm, you will grasp what I surmise since you were keen to hear it. Whether it is true or not only the god knows, but this is how I see it, namely that in the intelligible world the Form of the Good is the last to be seen, and with difficulty; when seen it must be reckoned to be for all the cause of all that is right and beautiful, to have produced in the visible world both light and the fount of light, while in the intelligible world it is itself that which produces and controls truth and intelligence, and he who is to act intelligently in public or in private must see it. — I share your thought as far as I am able.

Come then, share with me this thought also: do not be surprised that those who have reached this point are unwilling to occupy themselves with human affairs, and that their souls are always pressing upward to spend their time there, for this is natural if things are as our parable indicates. — That is very likely.

Further, I said, do you think it at all surprising that anyone coming to the evils of human life from the contemplation of the divine behaves awkwardly and appears very ridiculous while his eyes are still dazzled and before he is sufficiently adjusted to the darkness around him, if he is compelled to contend in court or some other place about the shadows of justice or the objects of which they are shadows, and to carry through the contest about these in the way these things are understood by those who have never seen Justice itself? — That is not surprising at all.

Anyone with intelligence, I said, would remember that the eyes may be confused in two ways and from two causes, coming from light into darkness as well as from darkness into light. Realizing that the same applies to the soul, whenever he sees a soul disturbed and unable to see something, he will not laugh mindlessly but will consider whether it has come from a brighter life and is dimmed because unadjusted, or has come from greater ignorance into greater light and is filled with a brighter dazzlement. The former he would declare happy in its life and experience, the latter he would pity, and if he should wish to laugh at it, his laughter would be less ridiculous than if he laughed at a soul that has come from the light above. — What you say is very reasonable.

We must then, I said, if these things are true, think something like this about them, namely that education is not what some declare it to be; they say that knowledge is not present in the soul and that they put it in, like putting sight into blind eyes. — They surely say that.

Our present argument shows, I said, that the capacity to learn and the organ with which to do so are present in every person's soul. It is as if it were not possible to turn the eye from darkness to light without turning the whole body; so one must turn one's whole soul from the world of becoming until it can endure to contemplate reality, and the brightest of realities, which we say is the Good. — Yes.

Education then is the art of doing this very thing, this turning around, the knowledge of how the soul can most easily and most effectively be turned around; it is not the art of putting the capacity of sight into the soul; the soul possesses that already but it is not turned the right way or looking where it should. This is what education has to deal with. — That seems likely.

Now the other so-called virtues of the soul seem to be very close to those of the body — they really do not exist before and are added later by habit and practice — but the virtue of intelligence belongs above all to something more divine, it seems, which never loses its capacity but, according to which way it is turned, becomes useful and beneficial or useless and harmful. Have you never noticed in men who are said to be wicked but clever, how sharply their little soul looks into things to which it turns its attention? Its capacity for sight is not inferior, but it is compelled to serve evil ends, so that the more sharply it looks the more evils it works. — Quite so.

Yet if a soul of this kind had been hammered at from childhood and those excrescences had been knocked off it which belong to the world of becoming and have been fastened upon it by feasting, gluttony, and similar pleasures, and which like leaden weights draw the soul to look downward – if, being rid of these, it turned to look at things that are true, then the same soul of the same man would see these just as sharply as it now sees the image towards which it is directed. – That seems likely.

End notes

1 These shadows of themselves and each other are never mentioned again. A Platonic myth or parable, like a Homeric simile, is often elaborated in considerable detail. These contribute to the vividness of the picture but often have no other function, and it is a mistake to look for any symbolic meaning in them. It is the general picture that matters.

2 *Odyssey* 11, 489-90, where Achilles says to Odysseus, on the latter's visit to the underworld, that he would rather be a servant to a poor man on earth than king among the dead.

2 Six Types of Love

Eric Fromm

BROTHERLY LOVE

The most fundamental kind of love, which underlines all types of love, is *brotherly love*. By this I mean the sense of responsibility, care, respect, knowledge of any other human being, the wish to further his life. This is the kind of love the Bible speaks of when it says: love thy neighbor as thyself. Brotherly love is love for all human beings; it is characterized by its very lack of exclusiveness. If I have developed the capacity for love, then I cannot help loving my brothers. In brotherly love there is the experience of union with all men, of human solidarity, of human atonement. Brotherly love is based on the experience that we are all one…

Brotherly love is love between equals: but, indeed, even as equals we are not always "equal"; inasmuch as we are human, we are all in need of help.

MOTHERLY LOVE

Motherly love by its very nature is unconditional. Mother loves the newborn infant because it is her child, not because the child has fulfilled any specific expectation.

Motherly love…is unconditional affirmation of the child's life and his needs. …Affirmation of the child's life has two aspects; one is the care and responsibility absolutely necessary for the preservation of the child's life and his growth. The other aspect goes further than mere preservation. It is the attitude which instills in the child a love for living, which gives him the feeling: it is good to be alive, it is good to be a little boy or girl, it is good to be on the earth!

But the child must grow. It must emerge from mother's womb, from mother's breast; it must eventually become a completely separate human being. The very essence of motherly love is to care for the child's growth, and the means to want the child's separation from herself. Here lies the basic difference to erotic love. In erotic love, two people who were separate become one. In motherly love, two people who were one become separate. The mother must not only tolerate, she must wish and support the child's separation. It is only at this stage that motherly love becomes such a difficult task, that requires unselfishness, the ability to give everything and to want nothing but the happiness of the loved one. It is also at this stage that many mothers fail in their task of motherly love. The narcissistic, the domineering, the possessive woman can succeed in being a "loving" mother as long as the child is small. Only the really loving woman, the woman who is happier in giving than taking, who is firmly rooted in her own existence, can be a loving mother when the child is in the process of separation.

FATHERLY LOVE

Fatherly love is conditional love. Its principle is "I love you *because* you fulfill my expectations, because you do your duty, because you are like me." In conditional fatherly love we find...negative and positive aspects. The negative aspect is the very fact that fatherly love has to be deserved, that it can be lost if one does not do what is expected. In the nature of fatherly love lies the fact that obedience becomes the main virtue, that disobedience is the main sin-and its punishment the withdrawal of fatherly love. The positive side is equally important. Since his love is conditioned, I can do something to acquire it, I can work for it; his love is not outside of my control as motherly love is.

The mother's and the father's attitudes toward the child correspond to the child's own needs. The infant needs mother's unconditional love and care psychologically as well as physically. The child, after six, begins to need father's love, his authority and guidance. Mother has the function of making him secure in life, father has the function of teaching him, guiding him to cope with those problems with which the particular society the child has been born into confronts him.

EROTIC LOVE

Brotherly love is love among equals; motherly love is love for the helpless. Different as they are from each other, they have in common that they are by their very nature not restricted to one person. If I love my brother, I love all my brothers; if I love my child, I love all my children; no, beyond that, I love all children, all that are in need of my help. In contrast to both types of love is *erotic love*; it is the craving for complete fusion, for union with one other person. It is by its very nature exclusive and not universal; it is also perhaps the most deceptive form of love there is.

First of all, it is often confused with the explosive experience of "falling" in love, the sudden collapse of the barriers which existed until that moment between two strangers. But...this experience of sudden intimacy is by its very nature short-lived. After the stranger has become an intimately known person there are no more barriers to be overcome, there is no more sudden closeness to be achieved. ...For them intimacy is established primarily through sexual contact. Since they experience the separateness of the other person primarily as physical separateness, physical union means overcoming separateness.

Beyond that, there are other factors which to many people denote the overcoming of separateness. To speak of one's own personal life, one's hopes and anxieties, to show oneself with one's childlike or childish aspects, to establish a common interest vis-a-vis the world—all this is taken as overcoming separateness. Even to show one's anger, one's hate, one's complete lack of inhibition is taken for intimacy. ...But all these types of closeness tend to become reduced more and more as time goes on. The consequence is one seeks love with a new person, with a new stranger. Again the stranger is transformed into an "intimate" person, again the experience of falling in love is exhilarating and intense, and again it slowly becomes less and less intense, and ends in the wish for a new conquest, a new love—always with the illusion that the new love will be different from the earlier ones. These illusions are greatly helped by the deceptive character of sexual desire.

Sexual desire aims at fusion—and is by no means only a physical appetite, the relief of a painful tension. But sexual desire can be stimulated by the anxiety of aloneness, by the wish to conquer or be conquered, by vanity, by the wish to hurt and even to destroy, as much as it can be stimulated by love. It seems that sexual desire can easily blend with and be stimulated by any strong emotion, of which love is only one. Because sexual desire is in the minds of most people coupled with the idea of love, they are easily misled to conclude that they love each other when they want each other physically. Love can inspire the wish for sexual union; in this case the physical relationship is lacking in greediness, in a wish to conquer or to be conquered, but is blended with tenderness. If the desire for physical union is not stimulated by love, if erotic love is not also brotherly love, it never leads to union in more than an orgiastic, transitory sense. Sexual attraction creates, for the moment, the illusion of union, yet without love this "love" leaves strangers as far apart as they were before...

In erotic love there is an exclusiveness which is lacking in brotherly love and motherly love. ...Frequently the exclusiveness of erotic love is misinterpreted as meaning possessive attachment. ...Erotic love is exclusive, but it loves in the other person all of mankind, all that is alive. It is exclusive only in the sense that I can fuse myself fully and intensely with one person only. Erotic love excludes the love for others only in the sense of erotic fusion, full commitment in all aspects of life—but not in the sense of deep brotherly love.

Erotic love, if it is love, has one premise. That I love from the essence of my being—and experience the other person in the essence of his or her being. In essence, all human beings are identical. We are all part of One; we are One. This being so, it should not make any difference whom we love. Love should be essentially an act of will, of decision to commit my life completely to that of one other person. This is, indeed, the rationale behind the idea of the insolubility of marriage. ...To love somebody is not just a strong feeling—it is a decision, it is a judgment, it is a promise. If love were only a feeling, there would be no basis for the promise to love each other forever. A feeling comes and it may go. How can I judge that it will stay forever, when my act does not involve judgment and decision?

Taking these views into account one may arrive at the position that love is exclusively an act of will and commitment, and that therefore fundamentally it does not matter who the two persons are. Whether the marriage was arranged by others, or the result of individual choice, once the marriage is concluded, the act of will should guarantee the continuation of love. This view seems to neglect the paradoxical character of human nature and of erotic love. We are all One—yet every one of us is a unique, unduplicable entity. ...Both views then, that of erotic love as completely individual attraction, unique between two specific persons, as well as the other view that erotic love is nothing but an act of will, are true—or, as it may be put more aptly, the truth is neither this nor that.

SELF-LOVE

It is widespread belief that, while it is virtuous to love others, it is sinful to love oneself...

Before we start the discussion of the psychological aspects of selfishness and self-love, the logical fallacy in the notion that love for others and love for oneself are mutually exclusive should be stressed. If it is a virtue to love my neighbor as a human being, it must be a virtue-and not a vice-to love myself, since I am a human being too...

We have come to the basic psychological premises on which the conclusions of our argument are built. Generally, these premises are as follows: not only others, but we ourselves are the "object" of our feelings and attitudes; the attitudes towards others and toward ourselves, far from being contradictory, are basically *conjunctive*. With regard to the problem under discussion this means: love of others and love of ourselves are not alternatives. On the contrary, an attitude of love toward themselves will be found in all those who are capable of loving others. *Love*, in principle, *is indivisible as far as the connection between "objects" and one's own self is concerned.* Genuine love is an expression of productiveness and implies care, respect, responsibility and knowledge. It is not an "effect" in the sense of being affected by somebody, but an active striving for the growth and happiness of the loved person, rooted in one's own capacity to love.

To love somebody is the actualization and concentration of the power to love. The basic affirmation contained in love is directed toward the beloved person as an incarnation of essentially human qualities. Love on one person implies love of man as such. The kind of "division of labor," as William James calls it, by which one loves one's family but is without feeling for the "stranger," is a sign of a basic inability to love. Love of man is not, as is frequently supposed, an abstraction coming after the love for a specific person, but it is its premise, although genetically it is acquitted in loving specific individuals... *The affirmation of one's own life, happiness, growth, freedom is rooted in one's capacity to love*...If an individual is able to love productively, he loves himself too; if he can love *only* others, he cannot love at all...

The *selfish* person is interested only in himself, wants everything for himself, feels no pleasure in giving, but only in taking. ...He can see nothing but himself; he judges everyone and everything from its usefulness to him; he is basically unable to love. Does not this prove that concern for others and concern for oneself are unavoidable alternatives? This would be so if selfishness and self-love were identical. But that assumption is the very fallacy which has led to so many mistaken conclusions concerning our problem. *Selfishness and self-love, far from being identical, are actually opposites.* The selfish person does not love himself too much but too little; in fact he hates himself. This lack of fondness and care for himself, which is only one expression of his lack of productiveness, leaves him empty and frustrated. He is necessarily unhappy and anxiously concerned to snatch from life the satisfactions which he blocks himself from attaining. He seems to care too much for himself, but actually he only makes an unsuccessful attempt to cover up and compensate for his failure to care for his real self.

LOVE OF GOD

It has been stated above that the basis for our need to love lies in the experience of separateness and the resulting need to overcome the anxiety of separateness by the experience of union. The religious form of love, that which is called the love of God, is, psychologically speaking, not different. It springs from the need to overcome separateness and to achieve union...

In all theistic religions, whether they are polytheistic or monotheistic, God stands for the highest value, the most desirable good. Hence, the specific meaning of God depends on what is the most desirable for a person...

The truly religious person, if he follows the essence of the monotheistic idea, does not pray for anything, does not expect anything from God; he does not love God as a child loves his father or his mother; he has acquired the humility of sensing limitations, to the degree of knowing that he knows nothing about God. God becomes to him a symbol in which man, at an earlier stage of his evolution, has expressed the totality of that which man is striving for, the realm of the spiritual world, of love, truth and justice, and considers all of his life only valuable inasmuch as it gives him the chance to arrive at an ever fuller unfolding of his human powers-as the only reality that matters, as the only object of "ultimate concern"; and, eventually, he does not speak about God-nor even mention his name. To love God, if he were going to use this word, would mean, then, to long for the attainment of the full capacity to love, for the realization of that which "God" stands for in oneself...

Having spoken of the love of God, I want to make it clear that I myself do not think in terms of a theistic concept, and that to me the concept of God is only a historically conditioned one, in which man has expressed his experience of his higher powers, his longing for truth and for unity at a given historical period.

3 "Varieties of Love: What is Love?"

Alan Soble

> In Love...one misheard word can be tremendously important. If you tell someone you love them...you must be...certain that they have replied "I love you back" and not "I love your back."
>
> —Lemony Snicket, *Horseradish (110)*

WHAT IS LOVE?

The difficulty of answering this question is shown by the facts that many answers exist and debates about "genuine" love seem endless and unresolvable. One methodology we can apply in seeking an answer is this: if a claim or theory about loves implies that many cases of what we ordinarily call love aren't love, that is a reason (not necessarily decisive) for denying the claim or theory. For example, some people think love lasts forever, that it endures as long as both lover and beloved are alive. However, suppose *X* and *Y* love each other during their senior year of college, but after graduation find jobs in different cities. They live too far apart to continue to nourish their love, which fades away. *X* and *Y*, by moving apart, decided to end their love. Are we not reluctant to say that merely because their love ended, it had never been love? As much as we hope that love lasts forever, we realize that many loves do end. Our methodology tells us to reject the "forever" thesis.

Similarly, consider the claims that if *X* loves *Y*, *X*'s actions always benefit only *Y*, or that if *X* loves *Y*, *X* could not love *Z* at the same time. If these claims are true, the world has seen very little love. Our methodology also implies that if a claim or theory of love implies that what we do not ordinarily call love is actually love, that is a strike against the claim. This principle allows us to deny that lust and infatuation, no matter how powerful, are genuine cases of love.

Love is similar to other complex concepts: "work of art" is a good example. What love and what art are, is equally perplexing. We label many things "art," applying that term to many noncontroversial cases. But in other cases applying "art" is debatable. The same holds for "love"; there are uncontroversial and debatable cases. By contrast, "automobile" possesses a simplicity that "art" and "love" lack. Perhaps this is because making a mistake about what is or is not an automobile is not momentous, certainly not as significant as confusing love with lust or infatuation. Emotions have a profound effect on our lives. Mistakes in love are as significant, if we are art investors, as confusing masterpieces with garbage.

Disagreements about how to apply "love" to specific cases might occur because love refers not to one thing but many. The psychologist Leo Buscaglia, however, says that "there are not 'kinds' of love. Love is of one kind. Love is love." Buscaglia means that the phenomena we call "love" (loving one's mother, loving chocolate, loving God, romantic love) all have something in common in virtue of which they are cases of love. All possess the "fine gold thread" of love. A problem with this view is that identifying the common ingredient among all these loves is difficult. If the common ingredient is that the lover cares for the well-being of the beloved, it would be nonsense to say that a child loves its mother. Children do not yet feel benevolence toward their parents; some loves contain more need than concern. If forming a unitive bond with one's beloved is the common ingredient, a child might have a desire to merge (or remain merged) with its mother, and so love her; but to say that the child loves (needs) its milk, so the child wants to merge with it, severely stretches the notion of merging.

The many kinds of love humans experience can be categorized in several ways. We can distinguish loves in terms of their *objects*, those items of the world that receive our love. We love our spouses, parents, siblings, children, friends, pets, country, abstract ideals, God, geometrical theorems, personal belongings, food, movies, books, and nature. Or loves can be differentiated by their *basis*, that which accounts for love's existence. (Later in this chapter we'll categorize loves as "eros-type" and "agape-type.") Loves can also be differentiated by their typical causal effects, whether a love causes a desire to benefit the beloved, or causes a desire for union with the beloved, or causes a desire to possess the beloved.

Some languages mark different loves by using different nouns. For example, the ancients distinguished loves linguistically, philosophically, and theologically using the Greek terms *eros* for erotic, passionate, desiring love; *agape* for Christian giving love; and *philia* for friendship-love. In Hungarian, *szerelem* means romantic, erotic love, while *szeretet* means parental love/brotherly love/God's love. But unfortunately, English uses the word "love" for all these phenomena, employing adjectives instead to mark differences among them. This linguistic fact about English acutely raises the question whether a common ingredient exists in all these things we call love.

LOVE AND VALUE

We commonly both want something and like it. Usually, we want it because we like it, instead of liking it *because* we want it. Occasionally, however, we like something just because we want it; these cases are psychologically interesting and puzzling. A similar correlation exists between loving something and finding value in it. Do we love something because we find value in it, or do we find value in it because we love it? Thus the relationship between loving persons and the value they have can be understood two ways: (1) *X* loves *Y* because *X* finds the features of *Y* valuable or attractive, and (2) *X* finds the properties of *Y* to be valuable, or *X* finds value in *Y*, because *X* loves *Y*. Similarly, we can ask whether *X* hates *Y* because *Y* has an annoying property, or *X* finds the properties of *Y* annoying because *X* hates *Y*. Indeed, about many emotions (fear, anger, admiration) we can ask whether *X* has the emotion toward *Y* because *Y* has some property *P*, or *X* believes *Y* has *P* because *X* experiences the emotion toward *Y*.

Loves that exhibit relation (1) between love and value will be called "eros-type" ("e-type"), while loves that exhibit relation (2) between value and love will be called "agape-type" ("a-type"). The theory of eros-type love derives from Plato's doctrine of *eros* in the *Symposium*, according to which *X* loves *Y* in virtue of *Y*'s beauty and goodness. The theory of agape-type love derives from Christian *agape*, God's or Jesus's love for humans in the New Testament. *Philia*, the word Aristotle used for friendship-love in *Nicomachean Ethics*, is a special kind of eros-type love that includes care for the beloved for the beloved's sake (as in a-type love), but is based on the moral and intellectual virtue of the beloved (hence is e-type).

In eros-type love, the value of *Y*, the object of love, exists first. Then *X*'s love for *Y* arises as a response to this antecedent value. (Aristotle: "No one falls in love who has not first derived pleasure from the looks of the beloved.") Love is the dependent variable; it is explained by the fact that *Y* possesses valuable properties. What is not explained in this theory is the existence of *Y*'s value, which is the independent variable requiring explanation outside the theory of e-type love.

Note that if *X* loves *Y* because *Y* has a valuable set of properties *P*, we can expect *X* to love someone else (say, Z) who also exhibits *P*. Plato rejoiced in this implication, calling it "great folly" for *X* not to love other people who also possess *P*. With so much beauty on the landscape, there is no reason to focus on a single instance to the exclusion of the others. The philosopher Robert Brown does not rejoice. For him, to love someone is "to value the person...as irreplaceable," so he rejects the theory that *X*'s love is based on *Y*'s valuable properties.

In agape-type love, the love exists first. Then the value of the beloved, as far as it figures into love, comes into existence. The beloved's value for the lover is the dependent variable, and its existence is explained by *X* loving *Y*. What is not explained in this theory is the love itself, the independent variable requiring explanation outside the theory of a-type love.

Several bases for a-type love will be mentioned later. For now we need to distinguish a-type from a "mixed" case that is really e-type. Suppose that some value V_1 in *Y* exists first, and *X* loves *Y* because of V_1. The love is based on, explained by, V_1. However, in virtue of loving *Y*, *X* attributes other values, V_2, to *Y*. This aspect of *X*'s love for *Y* is a-type. But *X*'s love is essentially e-type, because *X* attributing V_2 to *Y* and *X*'s love for *Y* are both explained by *X*'s response to V_1. These cases would be a-type only if *X*'s love was not grounded in V_1 or other valuable properties of *Y*.

EROS AND AGAPE

The agapic relation between *X*'s love for *Y* and *X*'s evaluation of *Y* is commonly asserted. For example, the philosopher Robert Kraut says that "someone's love for a specific person is not 'based upon' the belief that the person has superb qualities; if anything, it is the other way around." Kraut offers no argument, except a defense by observation: "Walter might judge Sandra to be the most marvellous person in the world, and these judgments might evoke feelings. But it seems to work precisely the other way around. The amorous feelings often come first; the favorable judgments...are already 'guided' by—that is, are a consequence of—the emotional responses." The philosopher W. G. MacLagan (1903-72) also claims that "*X* loves *Y* because *Y* has valuable *P*" is backwards, citing F H. Bradley's (1846-1924) *Aphorisms*: "We may approve of what we love, but we cannot love because we approve. Approbation is for the type, for what is common and therefore uninteresting."

The argument here is that e-type love is incompatible with love's being exclusive for one beloved. Were *X* to love *Y* because *X* approved of *Y*, *X* would have reason to love any *Z* of whom *X* approves. Approbation is for properties shared by many people; it is directed at *types* of people, while love is for the ineffably unique individual.

Recently, the philosopher Harry Frankfurt similarly weighed in on the side of a-type love:

> It is true that the beloved invariably *is*, indeed, valuable to the lover. However, perceiving that value is not at all an indispensable *formative* or *grounding* condition of the love. It need not be a perception of value in what he loves that moves the lover to love it. The truly essential relationship between love and the value of the beloved goes in the opposite direction. It is not necessarily as a *result* of recognizing their value and of being captivated by it that we love things [and people]. Rather, what we love necessarily *acquires* value for us *because* we love it. The lover does invariably and necessarily perceive the beloved as valuable, but the value he sees it to possess is a value that derives from and that depends upon his love.

This popular view suffers from various difficulties—which is not to say that the theory of e-type love has none (e.g., the exclusivity tangle). The philosopher Susan Wolf remarks about these views that both are troublesome. She finds the idea "that one should love what is worth loving...horribly wrong," but she finds "problematic" that "the question of whether something is *worthy* of our love...is out of place." Pope Benedict XVI (Joseph Ratzinger) is the optimist: *eros* and *agape* can "find a proper unity in the one reality of love."[16] We can have our cake and eat it.

If *X* loves *Y* because *Y* is beautiful or has good character, *X*'s love for *Y* is easily explained as an everyday emotional phenomenon. It would not differ from other e-type phenomena: *X* hating *Y* because *Y* has done something nasty to *X*, or *X* fearing a bear because the bear poses a threat to *X*'s life. Questions do remain about the source of *Y*'s beauty or goodness and why *X* responds to these properties and not others. These questions might (or not) have uncomplicated answers. By contrast, if *X* judges *Y* beautiful or good because *X* loves *Y*, we have answered the question of the source of *Y*'s value. But *X*'s love for *Y* remains unexplained.

Consider an extreme case. If *X* judges *Y* beautiful or good because *X* loves *Y*, despite evidence indicating that *Y* is a cheat and liar, the attribution of value to *Y* by *X* is mysterious, and explaining why *X* loves *Y* is difficult. The reason for *X*'s loving *Y* cannot be familiar but esoteric, because *X*'s love has a profound effect on *X*'s cognitive faculties. If *X* believes *Y* is beautiful or good because *X* loves *Y*, and other merits of *Y* do not explain *X*'s love, or if *X* believes that *Y* has done something nasty to *X* because *X* hates *Y*, and *Y* has not done something else nasty to *X* that grounds and warrants *X*'s hate, explaining *X*'s emotion (love or hate) is difficult. For this reason, the theory of e-type love looks appealing. That many people reject such reasoning and think that the a-type relation applies typically to love reveals their attitude toward love. They want to treat it as special, not comprehensible the way other, less lofty emotions are.

In Plato's *eros*, the beauty or goodness of an object (body, soul, mathematical proof, work of art, a philosophical insight) grounds our love for it. In other e-type loves, the attractive features of the beloved object also account for the love and determine its course. For example, in courtly love, the lover chooses "one woman as the exemplar of all significant virtues and [uses] that as the reason for loving her. ...The inherent excellence of her total personality... elicits his love." In erotic love, the attractiveness of the beloved obviously plays an important role, even if sexual desirability is subjective and some lovers are less discriminating than others.

Parental love, too, seems to be e-type: parents love a child not merely for the child's existence but because the child has the property "is a child of mine" ("contains a piece of me") which is, from the parent's perspective, meritorious. The parent's reason for loving the child is both general and selective. If this is why the parent loves child *X*, the parent has equal reason for loving child *Y*, yet has no reason for loving someone else's child. Parental love, then, is preferential in a way that God's *agape* and neighbor-love could not be. God loves all His children, but because everyone is one of His children, God's love is general but not selective. And God loves all His children equally, because all worldly meritorious properties are irrelevant to Him.

Again we stumble on a paradox. Is God's love for humans e-type, if God loves humans in virtue of their possessing the property "is a child, or creation, of mine"? If God loves humans at least in part because they contain a piece of Him (the soul), God is like the e-type lover Narcissus who falls in love with the beauty of his own reflection in the water. Yet, the paradigm case of a-type love is the Christian God's love for humans, in which the attractive and unattractive properties of humans (our values and defects) are irrelevant:

> God does not love that which is already...worthy of love, but on the contrary that which... has no worth acquires worth...by becoming the object of God's love. Agape has nothing to do with the kind of love that depends on the recognition of a valuable quality in its object; Agape does not recognize value but creates it.

Why does God love humans? What is the basis of His love? "The 'reason' why God loves men is that God is God, and this is reason enough." As the Swedish scholar Anders Nygren (1890-1978) wrote, "There is only one right answer. ...Because it is His nature to Love. ...The only ground for it is to be found in God himself."

That love is not based on the object's merit is true also of Christian neighbor-love, which demands that humans love each other as God loves humans. Its basis is obedience to God's love-commandment. It is patient, selfless, benevolent, and long-suffering. It never fails and endures all things. Hence Christians love the sinner, the stranger, the sick, the ugly, the enemy, as well as the righteous, one's kin, and oneself. Individual attractiveness plays no role in neighbor-love. As the Danish philosopher Søren Kierkegaard says about *agape*,

> The task is not: to find—the lovable object; but the task is: to find the object already given or chosen—lovable.

What should we choose to love, because we choose first and love second? Kierkegaard answers, "True love is precisely [to find] the unlovable object to be lovable." Kierkegaard opposes himself to "that simple wise man of ancient times," Socrates, who taught that the lover desires the beautiful and joked about loving the ugly, and to Aristotle, for whom love for the ugly or bad was impossible. Kierkegaard boldly announces that we *should* love "the ugly." Kierkegaard does not mean that *X* should love *Y* in virtue of *Y*'s ugliness if *X likes* ugliness. Rather, we should love those who are unlovable in the worldly sense of "lovable" exactly because they are unlovable, not because they are *per accidents* lovable in our subjective perceptions.

Frankfurt's answer to "whom should we love?" is different: a person should select one object for love instead of another because "it is *possible* for him to care about [love] the one and not about the other, or to care about the one in a way which is more important to him than the way in which it is possible for him to care about the other." Frankfurt's criterion is too weak; it rules out *only* objects that we cannot care about (logically? in virtue of natural laws? psychologically?). This advice is useless, for if *X* cannot care about *Y*, *X* certainly cannot choose to care about *Y*. However, Frankfurt's theory does alert us to the possible role of voluntary decision in our loving someone; it is not (always or only) a psychological reaction over which we have no control.

Neighbor-love can be interpreted not as a-type but as e-type love. If humans have value precisely because God has bestowed value on them, neighbor-love could be construed as a person's response to *this* value in another person. Or perhaps neighbor-love is a response to the valuable piece of God that exists in all humans. If so, neighbor-love is a universal or general e-type love (one that has, as its object, everyone). Here we should distinguish the claim (about love's *basis*) that in *X*'s neighbor-love for *Y*, *X* loves *Y* because *Y* has the valuable property "contains a piece of God," from the claim (about love's *object*) that in *X*'s neighbor-love for *Y* what *X* loves is not *Y per se* but that piece of God that dwells in *Y*. This complication about the basis and object of love arises also in Plato's doctrine of *eros* in the *Symposium*. Does *X* love *Y* because *Y* has valuable properties P, or does *X* love P itself, for which *Y* is merely the vehicle?

Agape, too, can be understood as e-type, at least to the philosopher J. Kellenberger, who claims that *agape* is a response to "the inherent worth of persons as persons." Whereas *eros* responds to the accidental value of a person, *agape* responds to a person's necessary (inherent) value, value not lost regardless of changes in the person's accidental values. If so, *agape* is a universal *eros*. But Kellenberger's view has odd consequences when applied to God's love for humans. It implies that God loves humans because of the value that He bestows on them as their creator. This yields a circle, if God bestows that value on humans because He loves them.

Although the human love for God is often seen as e-type, because we love God in virtue of His infinite magnificence, it can be understood as a-type. If we love God no matter what He does or what happens to us in His world, and continue to love Him with all our hearts, constantly, enduring all things, despite the bad we suffer in life that is ultimately due to Him as creator of everything, our love for God appears to be a-type. On the other hand, suppose we are able to love God, no matter what, only because we do not *believe* that the suffering that befalls us is genuinely bad—after all, God is good, and everything in His plan is for the best. If God never really does bad, loving God "no matter what" is vacuous. This love doesn't exhibit the unconditionality that persists in spite of genuine bad (as in God's *agape* for us, who sin). It is a love that persists because it has faith that no badness exists. So our love for God is e-type, after all; we love God because of His infinite goodness and the wisdom (albeit indecipherable) of His plan.

The philosopher C.D.C. Reeve claims that the Christian commandment, "love your neighbor as yourself" (Romans 13:9) is "problematic": "[I]f we do not know what turns on [causes] our love for others, we do not know what turns on our self-love either, and so do not know how to love them as we do ourselves." Does *X* knowing *how* to do something presuppose *X* knowing *why X* does it? *X* knowing how to love others, how to be benevolent toward them, does not depend on *X* knowing why *X* cares about their welfare. The commandment admonishes *X* to do well for others, just as *X*, without admonition, would do well for himself. That *X* has no clue why *X* cares about himself or extends himself to others does not prevent *X* from following that command. Note also the inference from "*X* does not know what makes *X* love *Y*" to "*X* does not know what makes *X* love *X*." That's a nonstarter, unless we assume that all that whatever is true of one type of love is also true of other types. However, we already know this is false.

Trying to expose what he believes is another Christian confusion, Reeve quotes a passage we have seen from Nygren's *Agape and Eros*: "God does not love that which is already... worthy of love, but on the contrary that which...has no worth acquires worth...by becoming the object of God's love." He remarks, "Theologically speaking, [this] is not a compelling idea." Why? "God cannot want value-conferring love [for himself, from humans], since he is already infinitely valuable." Because God does not want value-conferring [a-type] love, "he must want value-responsive [e-type] love [for himself, from humans]. Apparently, then, value-responsive love is better, since God surely wants the best....But that makes God seem a deficient lover—one who loves less well [in a value-conferring way] than he wants to be loved."

This is a cute paradox, but the argument contains several illuminating mistakes. It assumes that conferring value on *Y* necessarily means increasing *Y*'s value. That's false as a *general* account of what "conferring value" means. So we cannot conclude, as an instance, that conferring value on God increases God's value. Philosophers (including Frankfurt) who think love is the conferring or bestowal of value don't claim that the value of the object is necessarily objectively increased. That *X* makes something valuable for *X* does not necessarily increase its intrinsic value, and to increase its intrinsic value is not necessarily what *X* was trying to do when conferring value on it. Hence humans are not (yet) barred from loving

God by conferring "subjective" value on Him. Nor does it follow from the fact that when God confers value on humans He *does* increase their value, that humans who confer value on the things they love increase the value of those things. Perhaps there are differences between the way God confers value and the way humans do, differences rooted in the natures of God and humans.

Further, from the fact that God is perfect—has value that cannot be increased—it follows only that humans cannot *succeed* in conferring (objective) value on God. It does not follow that humans cannot love God in a value-conferring way, just because their love cannot achieve its goal. God might want humans to love Him in a futile but not, for that reason alone, disastrous way. That humans nobly *try* to confer value on God might be the ticket for Him. Why not say that humans love God in one way, erosically, and God loves humans another way, agapically?

But Reeve thinks that God's love for humans must be the same as the human love for God. He argues that if love is acting beneficially for the beloved, and God loves humans this way, then "our love for him [is] mysterious. For how can any activity of ours...benefit someone whose...perfection puts him beyond benefit?" But love need not succeed in benefiting the beloved to *be* love. The analysis of love's benevolent disposition should be counterfactual, for example, "*X* loves *Y* only if *X* would benefit *Y* if *X* could." A person having few resources who tries anyway to benefit another person may well exhibit this component of love. In a Kierkegaard parable, an old impoverished woman gives a penny as charity. She benefited no one, but her heart was full of love.

For God, no gap exists between the disposition and its fulfillment: "if *X* could" is always satisfied. There is a "mystery" only if the human love for God must be exactly the same as God's love for humans. It is unconvincing for Reeve to reply, "in the face of love's confusions, we are tempted to divide and conquer. Sexual love, parental love, fraternal love, friendship, agape, love of country, love of sports." Why not divide? Contrast *X*'s love for *X*'s child with *X*'s love for the adult person with whom *X* is having an affair or *X*'s love of books and music—all are different, phenomenologically and analytically.

EVALUATING AND ASSESSING LOVE

The philosopher John Brentlinger notes that *eros* "proceeds from" value and *agape* "creates" value. He suggests that the distinction between *eros* and *agape* derives from a logically prior difference between values being objective (they exist independently of the lover's attitudes and hence are the kind of thing love could "proceed from") and values being subjective (they exist in virtue of the lover's attitudes and hence are the kind of thing love could "create"). If so, we would have to resolve one of the perennial problems in metaethics: are values objective or subjective? However, values being objective or subjective does not correspond to the difference between e-type and a-type love, for in e-type love, *X* can love *Y* because *X* subjectively evaluates *Y* as beautiful or good. Brentlinger acknowledges this point and proposes that deciding between *eros* and *agape* reduces, instead, to deciding whether *X* believing *Y* to have value grounds *X*'s love or *X* loving *Y* leads *X* to bestow value on *Y*. At this point, he declares the problem unsolvable: "*Neither* concept is by itself sufficient to explain all cases of love.

Rather, some lovers and loves will be cases of *eros* and some of *agape*." We cannot conclude, on his view, that either type of love is "preferable." But there may be factors we can invoke in debating the merits of these loves: their morality and rationality.

Perhaps e-type love is preferable, because it encourages *X* to discover (by self-reflection) why *X* selects *Y* and not *Z* to love. Lovers are commonly called on to justify loving a particular person; even the beloved might press the lover with "why *me*?" *X* must provide *Y* (and the unloved *Z*) with a more convincing reason than "that's the way I feel" or "dunno, just do" (or "just don't"). Further, saying that no answer exists to the question whether e-type or a-type love is preferable is as inadequate as saying that no answer exists to the question whether e-type anger is preferable to a-type anger. If we are angry, we usually know why. Anger toward *Y* without reasons seems irrational, pathological, or morally blameworthy.

Proponents of a-type love think "it is better to love without a reason, than with a reason." Why? Perhaps reasonless a-type love is constant. Constancy, however, is not necessarily a virtue of love for lover or beloved. Some loves are too painful. And constancy, as already argued, might not be required for genuine love. Further, why might *X*'s inability to offer reasons for love secure any more constancy than e-type? Of course, if *X* loves *Y* for attractive properties *P*, *Y* losing some of *P* jeopardizes *X*'s love; e-type love is only as constant as its basis: *Y*'s value. But if *X* has *no* reasons for loving *Y*, constancy is in equal jeopardy. If love is not based on the beloved's merit, its constancy is due, presumably, to something about *X*'s nature, not to *Y*'s nature (the way God's love for humans is due to His nature, not ours).

The constancy of reasonless love might result from the lover's voluntary determination to benefit the beloved despite significant disagreements. Determination here is the basis of *X*'s a-type love. Or suppose love involves a promise *X* makes to *Y*; it is grounded in an act of will (the way God loves humans, by His will). Whether determination or will grounds a-type love, the same principle applies: love is only as constant as its basis. The constancy of a-type love would depend on the often unpredictable and uncontrollable vicissitudes of one's determination or will. Unlike God's will, our will and determination are imperfect.

Another consideration is the intentionality of love: emotions depend on beliefs and these beliefs need not be true. If I believe that my grandmother has died, I experience grief. This grief is real while I experience it, even if I later discover that she is alive (at which point the grief dissipates). In e-type love, *X* loves *Y* because *Y* has attractive features *P*. But we should characterize e-type love more precisely: *X* loves *Y* because *X believes* or *perceives* that *Y* has *P*. Indeed, e-type love exists even if *Y* does not have *P*, if *X* thinks that *Y* has *P*. If e-type love is intentional, then love can be based on the false belief that *Y* has *P*. And this love may be genuine, like grief based on the false belief that my grandmother has died. The intentionality (belief component) of a-type love is different. It is a consequence of love, not part of the antecedent reason for love (as in e-type).

If *X* loves *Y* because *X* believes that *Y* has *P*, whether *X*'s belief is true can influence our assessment of *X*'s love, *X*'s belief that *Y* has *P* might be caused by suspicious psychological processes (rationalization, projection, idealization). If so, *X*'s love-grounding belief is irrationally formed and likely false, *X*'s love is defective. At least, we expect "blind" love to dissipate as soon as *X* realizes the truth. Or maybe *X* has been culpably negligent about *Y*'s properties; *X* did insufficient research about *Y* to warrant believing that *Y* has *P*. Similarly, if *X* fears something because *X* believes it is dangerous, but *X*'s belief is founded on weak or no evidence, *X*'s fear is irrational. Or perhaps *X* has false beliefs because *Y* has deliberately deceived *X* into believing that *Y* has *P*. Here we blame *Y*, not *X*, judging *Y* immoral for the deception. People who exaggerate their merits, verbally or by their acts and appearance, may be attempting to induce love in another person. This tactic is often unsuccessful in the long run.

If *X* loves *Y* because *X* believes falsely that *Y* has *P*, is *X*'s emotion really love? Or if *X* does experience love, is it precisely *Y* that is the object of *X*'s love, or does *X* love only a figment of *X*'s imagination? For the philosopher W. Newton-Smith, if *X* loves *Y* falsely believing that *Y* has *P*, *X*'s emotion is love only if it continues after *X* corrects *X*'s beliefs. If *X*'s emotion ends, *X* "never loved anyone at all." Indeed, if *X* loves *Y* believing falsely that *Y* has *P* and later, after being corrected, still loves *Y*, *X*'s love is a-type.

Now consider hate. Suppose *X* believes that *Y* has done something vicious to *X* and *X* begins to hate *Y*. If *X*'s belief is false, should we deny that *X* experiences hate? Should we deny that *Y* is the object? No. We expect, if *X* corrects *X*'s beliefs about *Y*, *X*'s hate will dissipate. If it continues, it is irrational a-type hate. Still, some insist that if *X* loves *Y* because *X* believes falsely that *Y* has *P*, *X*'s emotion is not love. Or if *X* loves *Y* falsely believing that *Y* has *P*, the object of *X*'s love is *Y-qua-possessor-of-P*, who doesn't exist. Why not analyze these cases the same way, not treating love as different from hate or as something special?

Some philosophers argue provocatively that *X* having false beliefs (fantasies, projections) about *Y* is *required* for love. If we accurately knew our beloveds, we would be too repulsed to find them lovable. For the Slovenian psychoanalytic philosopher, Slavoj Žižek, sexual attraction, too, "needs some phantasmic screen. ...[A]ny contact with a 'real' flesh-and-blood other, any sexual pleasure that we find in touching another human being, is not something evident but something inherently traumatic, and can be sustained only in so far as this other enters the subject's fantasy-frame. ...What happens, then, when this screen dissolves? The act turns into ugliness—even horror."

A person can love another on the basis of almost any property found valuable or attractive; the theory of e-type love by itself places no limit on what particular *P* might elicit love. Perhaps we can make judgments about a person loving (or hating) someone on the basis of some properties rather than others. A logical "anything goes" does not entail a moral, pragmatic, or aesthetic "anything goes." To decide for which properties people should love others is difficult, and one can easily slide from philosophical inquiry into blandly moralizing. Several ideas, however, are worth considering.

In *Confessions* (4.13), Augustine wrote: "I was in love with beauty of a lower order and it was dragging me down. I used to ask my friends 'Do we love anything unless it is beautiful?... Unless there were beauty...in them, they would be powerless to win our hearts'." Augustine supposes that anything loved must be beautiful, which both Socrates and Diotima assert about *eros* in Plato's *Symposium* (201a, 202d, 203c, 204d, 205e, 209b). Yet Augustine proposes a distinction between beauty of "a lower order" (bodily beauty that arouses sexual desire) and beauty of a higher order (moral character), which is Diotima's strategy (*Symp.* 210a-d). Is love based on "higher" properties better than love based on "lower" properties?

Properties that ground love can be divided into two types: the superficial and significant. So a distinction can be made between loving someone for their superficial properties and loving them, in a superior way, for their significant properties. Suppose that a beautiful face, one's wealth, and one's power and position are superficial properties, while intelligence, grace, wit, honesty, and moral virtue are significant properties. (Pausanias says so in distinguishing between "vulgar" eros for the body and "heavenly" eros for the mind [*Symp.* 180d-181a, 183e]). *X* could love *Y* for her beautiful hair and sexual attractiveness (or for *Y*'s usefulness in helping *X* achieve *X*'s goals, as in Aristotle's use-friendship), but if *X* loved *Y* instead (or in addition) for *Y*'s integrity and courage (as in Aristotle's genuine friendship), *X*'s love would be superior. Why might loving for significant properties be superior? Loving someone for their beautiful hair or economic resources might show that the lover is shallow. Or perhaps loving for significant properties is more likely to be constant, showing it to be genuine, while love for superficial properties is unlikely to endure. (Is this true?) Or love based on significant properties reliably exhibits behaviors connected with love, say, being unselfishly concerned for the beloved's well-being. (Is this true?) Or loving for significant properties is loving a person for who he or she really is. Perhaps. But identifying properties as superficial and others as significant is philosophically difficult.

In this approach, the important thing is to come down on the right side in the battle between variable, unreliable physical attributes and more respectable psychological and ethical properties. We should love each other not for our firm body, but in spite of our flabby body and because of our attractive mind. But our character, too, is often ugly. Does our moral and intellectual beauty or goodness really attract others to us, overcoming their perception of our badness? Are we able to conceal the bad components of our character, making our mental flab invisible? If not, we would have to learn to love each other in spite of our having inadequate characters. But now, having ascended from loving the body to loving the mind or personality, to what do we ascend? Is it to the piece of God within us, or (recall Kellenberger's *agape*) to the "inherent worth of persons as persons"? Perhaps the only nonsuperficial value of a person is their "deep" value as a person. If so, we have lost all possibility of preferring to love one person instead of another, because everyone has this inherent value. Maybe we should admit, given the corruptness of our characters, that our fleeting, youthful physical beauty is our only beauty and the major cause of love. Love based on a beloved's superficial traits need not be a failure.

Consider another approach: love based on a lover's autonomously formed preferences is superior. What counts is not the type of value in the beloved but why the lover is drawn to some properties and not to others. The best love exists in virtue of the beloved's properties, whatever they are, that mesh with the lover's non-superficial preferences—preferences not unreflectively allowed to operate in the lover's choices.

The normative value of the preferential choice flows from the value of autonomy, not from the value of the object of that choice. For lovers to control adequately their preferences and distinguish worthy from unworthy objects of love is not due only to inner strength; social and economic conditions must encourage the free development and modification of preferences. If this requires the absence of economic and political inequality, social engineering may be necessary to refurbish not only sexuality but also love.

Loving because of autonomously formed preferences (or focusing on significant properties of beloveds, if you favor that view) is not the only requirement for avoiding defective love: a love may be morally inadequate if *X*'s belief that *Y* has *P* is epistemically weak. *X* loving *Y* on the basis of negligent beliefs may be morally wrong, for as a result of doing so *X* makes commitments to *Y*, creates in *Y* expectations about *X*'s future behavior, and leads *Y* to believe that *X* wants to spend time with *Y*, that *X* will share some experiences exclusively with *Y*, and that *Y* will be an object of *X*'s preferential concern. Creating these beliefs in *Y* may harm *Y*, because *X* will likely discover that *X*'s beliefs about *Y* are false, and *Y*'s expectations are dashed.

Pausanias says that it is "discreditable" for a pursued beloved to give in (sexually) too soon to a desirous lover (*Symp.* 184a5). The beloved must discern, with good evidence, whether the lover is motivated by vulgar or heavenly *eros* or whether the lover is feigning heavenly to hide the vulgar. If the pursued gives in too soon, he has only himself to blame. He has negligently allowed himself to be someone's beloved by not making sure he has true beliefs about his lover's character and motivation. Aristotle, too, thinks the beloved is blameworthy if he "has erroneously assumed that the affection he got was for his character, though nothing in his friend's conduct suggested anything of the sort."

Pausanias also says that if the beloved *Y* gives in to the lover *X* because *Y* believes that *X* is rich, *Y* is proceeding incorrectly, whether *Y*'s belief is true or false. In this case, *Y* is "disgraced"; he shows himself to be a shallow person attracted to superficial properties. On the other hand, if *Y* gives in to *X*, believing that *X* is wise and virtuous, "it does [*Y*] credit," even if the belief is false (*Symp.* 184e-185a). Maybe what Pausanias should say is that if *Y* believes falsely but on good evidence that *X* is wise, *Y* giving in to *X*'s pursuit is praiseworthy, because *Y* giving in reveals *Y*'s commendable motive (to become virtuous). But if *Y* believes falsely, due to negligence, that *X* is wise, then despite *Y*'s motivation *Y* is discredited. For *Y* has not seriously attempted to sort out heavenly from vulgar lovers.

THE FINE GOLD THREAD

Does love have a "central" ingredient (*C*)? *C* would be common to all loves, not only to all instances of one type of love but all types of love, period. *C* would not be something trivial. "Occurs in the Milky Way" is common to all cases of love we know about, but it does not help us understand love. Nor should we say that in every case of *X* loving something, *X* likes it. That might be true (oh?), but it is not illuminating. ("Why does *X* like *Y*?" is no more transparent than "Why does *X* love *Y*?") *C* should be able to explain why an occurrence or type of love *is* love. By referring to *C* we should be able to point out the difference between love and other things, all of which also occur in the Milky Way. *C* must provide a way to distinguish love not only from things that are wildly different from it (dogs, meatballs) but also from things that are close to love but are either pretenders or associated phenomena (infatuation, respect, admiration, sexual desire).

Because love is so varied—its types include vulgar sexual love, heavenly erotic love, romantic love, courtly love, neighborly love, parental love, and love for one's country and one's god and chocolate—it would be a magnificent feat to discover a significant common element in all cases. If some *C* is proposed as the "fine gold thread," one must argue that (1) *C* is found in all cases and types of love, including cases or types in which *C* intuitively seems absent *or* (2) cases of love in which *C* is absent are not cases of love, or at least not "true" cases. Alternatively, a weaker claim about the scope of *C* might be made: (3) *C*'s significance is restricted to one prototypical kind of love or only to those types of love that most strongly demand elucidation. The attempt to uncover the fine gold thread of love is abandoned, and we are asked to settle for the fine gold thread of some loves (i.e., the "important" ones). Or (4) we admit that love has no *C*. "Love," like "game," has no essence, but is a "family resemblance" concept.

The nature of this philosophical quest is nicely illustrated in Plato's account of love in the *Symposium*. Plato set out to discover the fine gold thread that was common, without exception, to all loves, including love for gods, persons, and objects. Plato's generic formula—love is the desire for the perpetual possession of the beautiful and good (*Symp*. 204b, 204e, 205d)—expresses the fine gold thread. This *C* expressed in the formula does illuminate love well by explaining some of its features. It explains why love is often, if not essentially, nonexclusive, directed at many persons and things. It explains love's inconstancy, switching from one object to another, abandoning the first for the second. Some earthly beauties deteriorate; they are inferior to other beauties that fade less quickly. Or *X*'s love for *Y* is inconstant because *X* realizes in advance that *Y*'s beauty *will* fade (*Symp*. 210b-c); *X*'s love dies even before *Y*'s inferior earthly beauty diminishes. Also, once *X* achieves satisfaction by possessing many instances of one type of beauty, *X* has had *X*'s fill. A higher level of happiness through love is attainable only by possessing higher beauties. And Plato's formula explains the burning, possessive passion of love (*Symp*. 204d-e): making the beautiful and good our own, part of ourselves, is a goal we pursue relentlessly, on pain of not fulfilling our potential or not achieving genuine happiness. For Plato, the exclusivity and constancy of love are eventually secured, but only when a person "ascends" to the highest stage of love's ladder. Here Absolute Beauty itself is contemplated, an unchanging, unique, infinite beauty, a beauty unadulterated by any earthly vehicle (*Symp*. 21 le). Beauty as beloved, though, unlike the Christian God, is an impersonal thing and does not reciprocate a human's love.

Plato suggests that all love (not only at the highest rung of the ladder) may be a desire for an inanimate thing. Goodness and beauty are properties—inanimate items—existing among the furniture of the universe. So, if one loves a beautiful horse, the horse's beauty is the "beloved," the object of one's possessive desire, not the horse. The horse is only the "occasion" of love, the vehicle that carries love's genuine object. This dimension of Plato's account of love holds up well when applied to a large number of things: love for the gods, art, and music, even sexual love in which a person is entranced by the beauty displayed by an arousing body. But Plato's account may have defects.

One criticism of Plato's view, raised originally and implicitly by Aristotle (about an inferior love, use-friendship, in which a person's utility is the object of "love"), and more recently and vigorously by Gregory Vlastos (1907-91), is that Plato's doctrine implies that *X* does not love *Y* as a person but rather loves only the goodness or beauty that *Y* manifests (features having an independent existence for which *Y* is only the vehicle). As Vlastos says, this is "the cardinal flaw in Plato's theory. It does not provide for love of whole persons, but only for love of that abstract version of persons which consists of the complex of their best qualities." The alternative superior love that Vlastos has in mind is a-type. We should strive for the unconditional acceptance of the other person. Only then is the "whole" person loved.

A Platonist can respond in three ways. First, Plato's view applies to personal love exactly as implied by his formula—*X* loves precisely the beauty and virtue *X* finds in *Y*. We should deflate our sentimental belief that in some delightful sense *X* loves *Y* "in (or for) himself" or "as a whole person." We must admit this even about happy loves, that we love a person's intelligence and courage but not "he himself." Second, the Platonist could say that personal love, because it fails to fit Plato's generic definition, is not genuine love but only a relative or pretender. This response is unappealing. Personal love is one of love's paradigm cases. Third, the Platonist could restrict the scope of Plato's definition, admitting that the formula does not apply to personal love (which *is* love) but only to a subset of the domain of love. This strategy is cowardly, eviscerating Plato's theory of *eros*. The first strategy is therefore the most ambitious.

Aristotle placed person-person love in a predominant position; he emphasized this by using the word *philia* instead of *eros* when talking about love. For Aristotle, Plato's focus on loving things was wrong because inanimate things cannot be beloveds:

> Love for a soulless thing is not called friendship, since there is no mutual loving, and you do not wish good to it. For it would presumably be ridiculous to wish good things to wine; the most you wish is its preservation so that you can have it. To a friend, however, it is said, you must wish goods for his own sake.

A thing cannot reciprocate the feelings or desires we have for it and, more significantly, we cannot, because of its dumb nature, wish it well for its own sake (*NE*, 1155b3 0). This phenomenon—*X* desiring for *Y* what is good for *Y*, and *X* pursuing *Y*'s good for *Y*'s benefit (not *X*'s)—is Aristotle's *C* of genuine love. But Aristotle did not intend this formula to be an umbrella definition covering all types of love. Rather, in emphasizing this "robust" concern, Aristotle was portraying personal love in its best or genuine form. Aristotle recognized two other types of love (use-friendship and pleasure-friendship), but didn't expect them to satisfy everything implied by his formula for perfect love.

Features of *eros*—the lack of exclusivity and constancy—reappear in Aristotle's *philia*. This is not surprising, because all three friendships are e-type; the basis is the friend's usefulness, pleasantness, or, in the best friendship-love, the other person's moral and intellectual virtue. For Aristotle, if a person loves another because he or she is pleasant or useful, this love is self-interested (*NE*, 1156all-17). These friendships will not be constant or exclusive; the basis of friendship (whether pleasant or useful) is widespread and does not endure (*NE*, 11156a20-1156b5, 1157aS-10).

Some of this is also true of Aristotle's best friendship. Aristotle does not expect friendship-love to be exclusive, even though persons of good character are rare (*NE*, 1156b25). One should, if one can, have many friends of good character. However, in contrast to inferior friendships based on usefulness, Aristotle expects constancy in ideal friendship. Goodness, as opposed to usefulness, usually endures, so love based on goodness endures (*NE*, H56b2). Nevertheless, Aristotle's best friendship, unlike a-type love but like e-type love, is conditional on the friend's remaining good (*NE*, 1165bl2-22). When Aristotle says that *X* should abandon *X*'s friendship with *Y* if *Y* has changed from being good to being evil (were that possible), he does not mean that the friendship ends for a self-interested reason. Rather, *X* realizes that *X* can no longer benefit *Y* for *Y*'s sake.

A few hundred years later, Paul and the authors of the New Testament gospels registered disagreement with Aristotle on the perfection of *philia*. Paul was dissatisfied because (like Plato's *eros*) Aristotle's *philia* was conditional on the beloved's merit. A person is drawn to another in response to their moral and intellectual virtue, even if the lover pursues the beloved's good for its own sake. Aristotle's *philia* is not perfect because it is benevolent only to those who are judged worthy of it; *philia* is not universal but preferential.

For Paul, though, perfect love graciously bestows value on its beloveds and does not respond to their unreliable value. Perfect love continues to love despite the beloved's faults, endlessly forgiving the sinner and trusting the enemy and the stranger, full of hope for the best (1 Corinthians 13). God's love is perfect. It is not *philia*, though. It is *agape*, which is shown in the sacrificial gift of His son for the sake of a humanity unworthy of such a blessing. Is a*gape*, then, the freely-bestowed creation of value by the lover toward the beloved, the only (true) love for the Christian tradition? If "yes," other types of love cannot be genuine. We cannot call *eros*, the human love for God, love. But if "no," *agape* is not the only (true) love, then the unconditional, creative bestowal of value is not the fine gold thread.

CONCERN AND BENEVOLENCE

Notice some other differences between the two types of love. An *eros*-type love is a response to a person's value, and might be largely independent of our will. We have feelings for the beloved, elicited by the value we perceive, and have little control over this process beyond attempting to form our own preferences. As a result, being ordered, by ourselves or others, to love someone seems incomprehensible. By contrast, being commanded to exhibit neighbor-love makes sense. *Agape*-type love, if it has any associated feelings, is not to the same extent as e-type love constituted by feelings. So if neighbor-love were constituted primarily by benevolence, that would explain why it can be commanded. We can be commanded to benefit other people even at expense to ourselves. And benefiting them (loving them) need not rest on their having value.

Perhaps, then, we should emphasize not *agape's* bestowal of value but its benevolence, its works of love (to use Kierkegaard's expression), acts done for the other's sake without thought of the self. If we pay attention to this aspect of *agape*, Aristotle (from a Christian perspective) did not make a huge mistake. He asserted, with Plato, that benevolence should be distributed according to antecedent merit, but he improved on Plato by seeing that caring about the beloved's welfare for the beloved's sake was essential to love at its best. Even

Plato recognized that love motivated people to perform virtuous acts. Phaedrus's contribution to the *Symposium*, in which he praises the power of love to induce brave and noble self-sacrificial deeds (178d-179c), as well as Diotima's teaching Socrates that the main effect of love was the begetting of beauty and goodness (*Symp*. 206b-e), show that Plato's eros is not selfishly self-centered (even if it aims ultimately at the lover's happiness). Plato, Aristotle, and Paul all maintain that benevolence is a feature of love. Perhaps, then, benevolence is love's fine gold thread. But even if benevolence is not *C* (at least because benevolence is absent from a child's love for its parents), concern for the welfare of the beloved is plausibly a necessary feature of personal love.

Psychologists Donn Byrne and Sarah Mumen opine that we should "learn to treat loved ones with as much politeness and kindness as we do strangers." This doesn't seem right. If we treat loved ones the way we treat strangers (drivers on the crowded highway; the elderly person next door whom we ignore), we would treat our beloveds with indifference. Byrne and Murnen admonish us to be nicer to our beloveds, as if our love for them was not expressed naturally as concern but destruction. Anyone who needs to be told to be nice to a beloved, however, is not a lover, or not a very good lover. (Are spouses or partners who abuse their beloveds less than ideal lovers or not lovers at all?) It seems that if *X* loves *Y*, *X* wants to benefit *Y*, and *Y* in particular. If *X* does not love *Y*, *X* is not expected to be especially concerned for *Y*'s welfare (unless we construct elaborate or intricate stories to explain it). Perhaps *X*'s desire to benefit *Y* can be conceived as a psychological effect of love. But because saying that *X* loves *Y*, but *X* does not care about the happiness or flourishing of *Y*, makes little sense, the link between *X* loving *Y* and *X* wanting to benefit *Y* might be tighter: wanting to benefit the beloved might partially constitute love.

Suppose, then, that if *X* loves *Y*, *X* wants the good for *Y* and acts, as far as *X* is able, to pursue *Y*'s good. But to what extent is *X* concerned to help *Y* flourish? How much or how often is *X* prepared to sacrificed *X*'s good for *Y*? This issue arises in arguments between two persons, both of whom claim to love the other. One says, "if you loved me, you would do this act," and the other replies, "if you loved me, you wouldn't ask me to do it." They disagree about the extent or nature of the beneficial acts that are required by love. Who is right? Will that depend on what the act is (say, a sexual act), or is that irrelevant? These questions are interesting and important.

Several things might be said about love's benevolence without fanfare. *X* loving *Y* is compatible with *X* not sacrificing *X*'s life to secure a less valuable good for *Y*, although never sacrificing anything for *Y* negates *X*'s claim to love *Y*. If *Y* desires only that *X*'s desires be satisfied, *X* will fully satisfy *Y* as soon as *X*'s desires are satisfied (but it would be strange to say that *X* has pursued *Y*'s good). *X*, as *Y*'s lover, will sometimes be divided between seeking for *Y* what *Y* wants and seeking something else, especially what *X* thinks is good for *Y*. A conflict may thus arise between *X*'s concern for *Y* and *Y*'s autonomy.

The lover's abiding by the beloved's autonomous desires may jeopardize *Y*'s well-being by foregoing what is good for *Y* (even if *Y* protests otherwise); and to pursue these other goods might paternalistically deny *Y*'s personhood (violating Kantian ethics). Further, conflicts often occur between love and morality (or justice). Singling out one's beloved to be the recipient of special, preferential concern might violate moral standards of equality. If deciding how much of *X*'s good *X* must be willing to sacrifice for the sake of *Y* is difficult, so, too, is deciding how much of *Z*'s good *X* must *not* jettison in pursuing *Y*'s welfare, where *Z* might be a relative (parent, child) or needy stranger.

X might hate *Y* even if *X* respects and finds *Y* attractive. Hating the beautiful and good, which are ordinarily lovable, is possible. Indeed, one common response to successful people is to envy or resent them rather than admire or love them. Our reaction to a beautiful, witty, friendly, and educated person might be dislike. Similarly, instead of experiencing hate or anger, *X* might love a *Y* who abuses *X*; loving a despicable *Y* is also possible. Further, even though *X* would usually hate *Y* if *Y* has harmed *X*, sometimes *X* hates *Y* because *Y* has benefited *X*. Loving behavior can generate not return love or gratitude but annoyance.

These observations prompt the question of how to distinguish love from hate. Love and hate can be inspired by the same properties as well as accompanied by similar acts toward their objects. We are sometimes cruel to our beloveds; this mistreatment is indistinguishable from the harm done to a person we hate. The desire to be in the presence of the beloved may be a component of love, but it is also found in a perverse hate in which *X* relishes opportunities to be with *Y* to provoke *Y*. What is the difference? Perhaps love and hate are distinguishable only by their phenomenal feels, by how they register in the consciousness of the person experiencing them. Then the only person who can distinguish *X*'s love from *X*'s hate is *X*, for only *X* has access to *X*'s conscious states. But how did *X* learn how to label these internal states, if not by watching the external behavior of other people? We might rely on the theory of a-type love to differentiate love and hate: when *X* loves *Y*, *X* is caused to evaluate *Y* positively; when *X* hates *Y*, *X* is caused to evaluate *Y* negatively. This will not work for e-type love; it works for a-type love and hate only because we have no clear explanation why a-type love and hate arise in the first place.

One way of identifying emotions is to refer to their characteristic beliefs: *X* experiences fear when *X* believes *Y* is dangerous; *X* experiences gratitude when *X* believes that *Y* has benefitted *X*. Yet if *X* can hate or love *Y* when *X* believes that *Y* has the same P, this attempt to distinguish love and hate apparently fails. Accordingly, philosopher D. Hamlyn claims that love and hate are "not differentiated by beliefs [but] by factors other than beliefs." Nevertheless, if we want to rely on *X*'s beliefs about the emotion's object, the distinction might be this: love occurs if the attractive properties outweigh or make insignificant the unattractive properties of its object, while in hate the relationship is the other way around. In other words, e-type love and hate would depend on the interplay between the attractiveness and unattractiveness of their objects. (I'm not sure this solution works.)

Robert Brown has suggested, alternatively, that

> If we are to distinguish love from hate, the former must embody recognizable good will toward the beloved and the latter ill-will toward the victim. A lover who over the long term wished only ill-will toward his consort would be as much a definitional absurdity as a man who, filled with hate for his consort, would wish her only good.

Hate with no ill-will and only goodwill is not hate; love with no goodwill and only ill-will is not love. Love and hate can be distinguished by the desires and actions that follow from them—desires that constitute or are the effects of love and hate. But Brown's proposal provides only a weak criterion: no goodwill means no love. This account adequately handles the case of the lover who occasionally hurts the beloved. This person at least shows some goodwill toward the beloved, and counts as a lover. But a stronger criterion seems required: as long as the pattern of *X*'s desires over the long haul is mostly a string of goodwill desires, *X* loves *Y*; if the pattern is mostly a string of ill-will desires, *X* does not love *Y*, and might in fact hate *Y*.

Is "mostly" here quantitative or qualitative? We have not come any closer to settling how much, and of what kind, *X* must be willing to give up for the sake of *Y*, if *X* loves *Y*.

UNION

Then there's the union theory of love, according to which *C* is a physical, psychological, or spiritual union between the lovers (or desire for union), in which they form a new entity, the *we*. The union theory goes back at least to the ancient Hebrews (Genesis 2: 23-24):

> The man said, "This is now bone of my bones and flesh of my flesh; she shall be called 'woman, for she was taken out of man." For this reason a man will leave his father and mother and be united to his wife, and they will become one flesh.

In the *Symposium*, Plato put a tragicomical union view into the mouth of Aristophanes, who tells a romantic story about two half-persons wanting to be welded together into the whole they had once been (*Symp.* 189c-193e). Diotima replies in the dialogue that Aristophanes union view of love is misleading: "Love is not desire either of the half or of the whole," she tells Socrates (consistently with the Platonic formula), "unless that half or whole [is] good" (*Symp.* 205e). Plato, for whom love is the desire to possess the beautiful and good, advanced a union view in the sense that possessing the beautiful and good involved incorporating that beauty or goodness into one's self.

Michel Montaigne, inspired by his friendship with Étienne de La Boétie (1530-63), offered a union view, claiming that in love, "each gives himself so entirely to his friend that he has nothing left to share with another" (not even one's spouse or children?). For G.W.F. Hegel (1770-1831), "love is indignant if part of the individual is severed and held back." More recently, theologians Paul Tillich (1886-1965) and Karol Wojtyla advanced union views; psychoanalytic philosopher Eric Fromm (1900-1980) argued that love, as union, was the solution to life's major problem (aloneness); psychiatrist Willard Gaylin made union the central theme of his book on love; and the analytic philosophers J.F.M. Hunter, Mark Fisher, and Robert Nozick have proclaimed that a union, fusion, or merging of two beings into one lies at the heart of love. Novelist Philip Roth dissents: "People think that falling in love they make themselves whole? The Platonic union of souls? I think otherwise. I think you're whole before you begin. And the love fractures you. You're whole, and then you're cracked open."

Suppose that concern for the well-being of the beloved is important in personal love, even if not the fine gold thread of love. Can the union theory of love allow that Aristotelian "robust" concern or the benevolence of *agape* are features of personal love?

Consider Hegel. Love "proper," he says, is "true union." "In love, life is present as a duplicate of itself and as a single and unified self." Of what does this union consist? "To say that the lovers have an independence...means *only* that they may die" separately. Hegelian lovers are unified except that, being physical creatures, they might not die at the same time; one might live, disunified, without the other. Hegel means this literally, for he asserts that "consciousness of a separate self disappears, and *all* distinction between the lovers is annulled."

Hegel's idea is not plausible about adult persons who love each other and normally retain consciousness of themselves as distinct persons. Of course, all human relationships involve some loss of independence. When a person becomes close to another, both gain beliefs from each other, or abandon beliefs. Both also gain and lose options for behavior. Reciprocal modeling of personality traits also occurs. This is ubiquitous. But Hegel's loss of consciousness of the self and of the other person as separate entities, or the annulment of all distinction between the lovers, makes it hard to fathom how one person in a union-love could be concerned for the welfare of the other for the other's *own* sake. Aristotelian robust concern seems in jeopardy.

Wojtyla tries to solve the problem. He argues that because two Is merge into a single *we* (the fused entity of love) we cannot "speak of selfishness in this context." If the persons remain separate, two distinct foci of interest or well-being exist, which means that selfishness is logically possible. But if two lovers merge into one entity, selfishness is logically eliminated, because a single entity cannot be selfish toward itself or treat itself selfishly. Hence, Wojtyla apparently implies that love as union, in ruling out selfishness, must therefore contain genuine concern.

The argument, however, supports the contrary thesis that the loss of independence in union-love is incompatible with robust concern for the other person. If a union of two people into a single entity eliminates their being selfish toward each other, it also eliminates their having robust concern for each other. Wishing the other well for *their* sake, just as does being selfish, presupposes that each person is an independent focus of well-being. If, as Hegel says, in union-love *X* and *Y* form a single, unified self, then in love only one entity acts on its own behalf. We might imagine a single entity or person wishing himself or herself well for its own sake, as one person might respect himself or herself. However, this concern aimed at one's self is not robust concern; it is only self-interest. Further, were *X* to (attempt to) sacrifice *X*'s good for *Y*'s sake, *X* could only be sacrificing the joint good of *X* and *Y* for *Y*'s sake, not *X*'s own good, because *X* no longer has any good of *X*'s own that *X* could sacrifice. Hence, the logical possibility of self-sacrifice is also eliminated. The well-beings of the lovers *not* being joined is logically necessary for *X* to exhibit either selfishness toward *Y* or robust or sacrificial concern for *Y*. Wojtyla expects Christian married couples to introduce "love into love," to infuse *agape* into their sexual love or change sexual love into *agape*. This is impossible if the spouses have become a unity that excludes robust concern.

As Fromm suggests, we do need to overcome our existential or metaphysical aloneness; without an intimate union with another person we often feel distressingly incomplete. We desire to join our life with the life of another person, to mingle or merge our selves with others' selves. Once the union has taken root we have a motive for nurturing the interpersonal connection that alleviates our angst. Yet, we also require a sense of ourself as a distinct individual, which depends on maintaining separateness from the other. We need to glory in the exercise of our own abilities and to be responsible in our own right for acting in the world in accordance with our natures. We can often (or partially) satisfy both needs at the same time or alternatively. Yet, resolving the tension between the need for closeness with another person and the need for independence is challenging.

Aristophanes cannot be right (he was joking) that what lovers want more than anything else is to be welded together. We undergo the strenuous process of separating ourselves from parents and other powerful significant others precisely to become our own authentic persons. Why forsake the fruit of that onerous and exhausting labor by establishing ourselves once again within a union? Why give up that hard-gained autonomy for a union-love, abandoning our prize? If women in our culture have more difficulty than men individuating themselves from others and achieve whatever autonomy they do achieve more precariously and painfully, why should they be willing to throw it away?

One answer is that women are not interested in autonomy in the first place. Autonomy is a masculine value and trait. The union theory, in emphasizing a merging of selves at the expense of autonomy, may be true to women's nature and experiences. Thus, criticizing the union theory by invoking the value of autonomy may be merely to rely on masculine values to deflate women's values.

Nevertheless, union-love may not be good for women after all. Although in union-love both people theoretically merge into each other and become a single *we*, in practice women have merged into men, losing their identities in the relationship while men maintain distance from it. The existentialist feminist philosopher Simone de Beauvoir, in expressing the union view, claims that "the supreme goal of human love...is identification with the loved one." Beauvoir proceeds to describe "the woman in love":

> [She] tries to see with his eyes....When she questions herself, it is his reply she tries to hear.... She uses his words, mimics his gestures.

This description suggests that a woman in love substantially loses her self. Perhaps were the man to merge equally with the woman, all would be well. But maybe the woman should be concerned, as the man is, to maintain her autonomy and independent identity.

4 *Nicomachean Ethics*: Self Love

Aristotle

11.5 Self-love is a component of Friendship
11.51 The common view identifies self-love with selfishness

There is also a puzzle about whether one ought to love oneself or someone else most of all; for those who like themselves most are criticized and denounced as self-lovers, as though this were something shameful.

Indeed, the base person does seem to go to every length for his own sake, and all the more the more vicious he is; hence he is accused, e.g., of doing nothing of his own accord. The decent person, on the contrary, acts for what is fine, all the more the better he is, and for his friend's sake, disregarding his own good.

11.52 But facts about friendship justify self-love

The facts, however, conflict with these claims, and that is not unreasonable.

For it is said that we must love most the friend who is most a friend; and one person is most a friend to another if he wishes goods to the other for the other's sake, even if no one will know about it. But these are features most of all of one's relation to oneself; and so too are all the other defining features of a friend, since we have said that all the features of friendship extend from oneself to others.

All the proverbs agree with this too, e.g. speaking of 'one soul', 'what friends have is common', 'equality is friendship' and 'the knee is closer than the shin'. For all these are true most of all in someone's relations with himself, since one is a friend to himself most of all. Hence he should also love himself most of all.

11.53 Hence we must distinguish good and bad forms of self love

It is not surprising that there is a puzzle about which view we ought to follow, since both inspire some confidence; hence we must presumably divide these sorts of arguments, and distinguish how far and in what ways those on each side are true.

Perhaps, then, it will become clear, if we grasp how those on each side understand self-love.

11.54 The bad form of self-love is selfish, resting on an incorrect view of the self

Those who make self-love a matter for reproach ascribe it to those who award the biggest share in money, honours and bodily pleasures to themselves. For these are the goods desired and eagerly pursued by the many on the assumption that they are best; and hence they are also contested.

Those who are greedy for these goods gratify their appetites and in general their feelings and the non-rational part of the soul; and since this is the character of the many, the application of the term ['self-love'] is desired from the most frequent [kind of self-love], which is base. This type of self-lover, then, is justifiably reproached.

And plainly it is the person who awards himself these goods whom the many habitually call a self-lover. For someone who is always eager to excel everyone in doing just or temperate actions or any others expressing the virtues, and in general always gains for himself what is fine, no one will call him a self-lover or blame him for it.

11.55 But the good form of self-love rests on a correct view of the self

However, it is this more than the other sort of person who seems to be a self-lover. At any rate he awards himself what is finest and best of all, and gratifies the most controlling part of himself, obeying it in everything. And just as a city and every other composite system seems to be above all its most controlling part, the same is true of a human being; hence someone loves himself most if he likes and gratifies this part.

Similarly, someone is called continent or incontinent because his understanding is or is not the master, on the assumption that this is what each person is. Moreover, his own voluntary actions seem above all to be those involving reason. Clearly, then, this, or this above all, is what each person is, and the decent person likes and gratifies this most of all.

Hence he most of all is a self-lover, but a different kind from the self-lover who is reproached, differing from him as much as the life guided by reason differs from the life guided by feelings, and as much as the desire for what is fine differs from the desire for what seems advantageous.

11.56 Hence it leads to virtuous action

Those who are unusually eager to do fine actions are welcomed and praised by everyone. And when everyone contends to achieve what is fine and strains to do the finest actions, everything that is right will be done for the common good, and each person individually will receive the greatest of goods, since that is the character of virtue.

Hence the good person must be a self-lover, since he will both help himself and benefit others by doing fine actions. But the vicious person must not love himself, since he will harm both himself and his neighbours by following his base feelings.

For the vicious person, then, the right actions conflict with those he does. The decent person, however, does the right actions, since every understanding chooses what is best for itself and the decent person obeys his understanding.

11.57 It even leads to costly sacrifices

Besides, it is true that, as they say, the excellent person labours for his friends and for his native country, and will die for them if he must; he will sacrifice money, honours and contested goods in general, in achieving what is fine for himself. For he will choose intense pleasure for a short time over mild pleasure for a long time; a year of living finely over many years of undistinguished life; and a single fine and great action over many small actions.

This is presumably true of one who dies for others; he does indeed choose something great and fine for himself. He is ready to sacrifice money as long as his friends profit; for the friends gain money, while he gains what is fine, and so he awards himself the greater good. He treats honours and offices the same way; for he will sacrifice them all for his friends, since this is fine and praiseworthy for him. It is not surprising, then, that he seems to be excellent, when he chooses what is fine at the cost of everything. It is also possible, however, to sacrifice actions to his friend, since it may be finer to be responsible for his friends's doing the action than to do it himself. In everything praiseworthy, then, the excellent person evidently awards himself more of what is fine.

In this way, then, we must be self-lovers, as we have said. But in the way the many are, we ought not to be.

5 Excerpts from: *Theogony*

Hesiod

INVOCATION OF THE MUSES

Begin our singing with the Helikonian Muses,
Who possess Mount Helikon, high and holy,
And near its violet-stained spring on petalsoft feet
Dance circling the altar of almighty Kronion,

And having bathed their silken skin in Permessos
Or in Horse Spring or the sacred creek Olmeios,
They begin their choral dance on Helikon's summit
So lovely it pangs, and with power in their steps
Ascend veiled and misted in palpable air
Treading the night, and in a voice beyond beauty
They chant:

> Zeus Aegisholder and his lady Hera
> Of Argos, in gold sandals striding,
> And the Aegisholder's girl, owl-eyed Athene,
> And Phoibos Apollo and arrowy Artemis,
> Poseidon earth-holder, earthquaking god,
> Modest Themis and Aphrodite, eyelashes curling,
> And Hebe goldcrowned and lovely Dione,
> Leto and Iapetos and Kronos, his mind bent,
> Eos and Helios and glowing Selene,
> Gaia, Okeanos, and the black one, Night,

And the whole eerie brood of the eternal Immortals.

And they once taught Hesiod the art of singing verse,
While he pastured his lambs on holy Helikon's slopes.
And this was the very first thing they told me,
The Olympian Muses, daughters of Zeus Aegisholder:

"Hillbillies and bellies, poor excuses for shepherds:
We know how to tell many believable lies,
But also, when we want to, how to speak the plain truth."

So spoke the daughters of great Zeus, mincing their words.
And they gave me a staff, a branch of good sappy laurel,
Plucking it off, spectacular. And they breathed into me
A voice divine, so I might celebrate past and future.
And they told me to hymn the generation of the eternal gods,
But always to sing of themselves, the Muses, first and last.

But why all this about oak tree or stone?

Start from the Muses: when they sing for Zeus Father
They thrill the great mind deep in Olympos
Telling what is, what will be, and what has been
Blending their voices, and weariless the sound
Flows sweet from their lips and spreads like lilies,
And Zeus' thundering halls shine with laughter,
And Olympos' snowy peaks and the halls of the gods
Echo the strains as their immortal chanting
Honors first the primordial generation of gods
Whom in the beginning Earth and Sky bore
And the divine benefactors born from them;
And, Second, Zeus, the Father of gods and men,
Mightiest of the gods and strongest by far;
And then the race of humans and of powerful Giants.
And Zeus' mind in Olympos is thrilled by the song
Of the Olympian Muses, the Storm King's daughters.

They were born of Pieria after our Father Kronion
Mingled with memory, who rules Eleutherae's hills.
She bore them to be a forgetting of troubles,
A pause in sorrow. For nine nights wise Zeus
Mingled with her love, ascending her sacred bed
In isolation from other Immortals.
But when the time drew near, and the seasons turned,

And the moons had waned, and the many days were done,
She bore nine daughters, all of one mind, with song
In their breasts, with hearts that never failed,
Near the topmost peak of snowcapped Olympos.

There are their polished dancing grounds, their fine halls,
And the Graces and Desires have their house close by,
Who serves the Muses chants the deeds of past men
Or the blessed gods who have their homes on Olympos,
He soon forgets his heartache, and of all his cares
He remembers none: the goddesses' gifts turn them aside.

Farewell Zeus's daughters, and bestow song that beguiles.
Make known the eerie brood of the eternal Immortals
Who were born of Earth and starry Sky,
And of dusky Night, and whom the salt Sea bore.
Tell how first the gods and earth came into being
And the rivers and the sea, endless and surging,
And the stars shining and the wide sky above;
How they divided wealth and allotted honors,
And first possessed deep-ridged Olympos.

Tell me these things, Olympian Muses,
From the beginning, and tell which of them came first.

THE FIRST GODS

In the beginning there was only Chaos, the Abyss,
But then Gaia, the Earth, came into being,
Her broad bosom the ever-firm foundation of all,
And Tartaros, dim in the underground depths,
And Eros, loveliest of all the Immortals, who
Makes their bodies (and men's bodies) go limp,
Mastering their minds and subduing their wills.

From the Abyss were born Erebos and dark Night.
And Night, pregnant after sweet intercourse
With Erebos, gave birth to Aether and Day.

Earth's first child was Ouranos, starry Heaven,
Just her size, a perfect fit on all sides.
And a firm foundation for the blessed gods.
And she bore the Mountains in long ranges, haunted

By the Nymphs who live in the deep mountain dells.
Then she gave birth to the barren, raging Sea
Without any sexual love. But later she slept with
Ouranos and bore Ocean with its deep currents,
And also: Koios, Krios, Hyperion, Iapetos,
Theia, Rheia, Themis, Mnemosyne,
Gold-crowned Phoibe and lovely Tethys.

THE CASTRATION OF OURANOS

After them she bore a most terrible child,
Kronos, her youngest, an arch-deceiver,
And this boy hated his lecherous father.

She bore the Cyclopes too, with hearts of stone,
Brontes, Steropes and ponderous Arges,
Who gave Zeus thunder and made the thunderbolt.
In every other respect they were just like gods,
But a lone eye lay in their foreheads' middle.
They were nicknamed Cyclopes because they had
A single goggle eye in their foreheads' middle.
Strong as the dickens, and they knew their craft.

And three other sons were born to Gaia and Ouranos,
Strong, hulking creatures that beggar description,
Kottos, Briareos, and Gyges, outrageous children.
A hundred hands stuck out of their shoulders,
Grotesque, and fifty heads grew on each stumpy neck.
These monsters exuded irresistible strength.
They were Gaia's most dreaded offspring,
And from the start their father feared and loathed them.
Ouranos used to stuff all of his children
Back into a hollow of Earth soon as they were born,
Keeping them from the light, an awful thing to do,
But Heaven did it, and was very pleased with himself.

Vast Earth groaned under the pressure inside,
And then she came up with a plan, a really wicked trick.
She created a new mineral, grey flint, and formed
A huge sickle from it and showed it to her dear boys.
And she rallied them with this bitter speech:

"Listen to me, children, and we might yet get even

With your criminal father for what he has done to us.
After all, he started this whole ugly business."

They were tongue-tied with fear when they heard this.
But Kronos, whose mind worked in strange ways,
Got his pluck up and found the words to answer her:

"I think I might be able to bring it off, Mother.
I can't stand Father; he doesn't even deserve the name.
And after all, he started this whole ugly business."

This response warmed the heart of vast Earth.
She hid young Kronos in an ambush and placed in his hands
The jagged sickle. Then she went over the whole plan with him.
And now on came great Ouranos, bringing Night with him.
And, longing for love, he settled himself all over Earth.
From his dark hiding-place, the son reached out
With his left hand, while with his right he swung
The fiendishly long and jagged sickle, pruning the genitals
Of his own father with one swoop and tossing them
Behind him, where they fell to no small effect.
Earth soaked up all the bloody drops that spurted out,
And as the seasons went by she gave birth to the Furies
And to great Giants gleaming in full armor, spears in hand,
And to the Meliai, as ash-tree nymphs are generally called.

THE BIRTH OF APHRODITE

The genitalia themselves, freshly cut with flint, were thrown
Clear of the mainland into the restless, white-capped sea,
Where they floated a long time. A white foam from the god-flesh
Collected around them, and in that foam a maiden developed
And grew. Her first approach to land was near holy Kythera,
And from there she floated on to the island of Kypros.
There she came ashore, an awesome, beautiful divinity.
Tender grass sprouted up under her slender feet.
Aphrodite
Is her name in speech human and divine, since it was in foam
She was nourished. But she is also called Kythereia since
She reached Kythera, and Kyprogenes because she was born
On the surf-line of Kypros, and Philommedes because she loves
The organs of sex, from which she made her epiphany.
Eros became her companion, and ravishing Desire waited on her

At her birth and when she made her debut among the Immortals.
From that moment on, among both gods and humans,
She has fulfilled the honored function that includes
Virginal sweet-talk, lovers' smiles and deceits
And all of the gentle pleasures of sex.

But great Ouranos used to call the sons he begot
Titans, a reproachful nickname, because he thought
They had over-reached themselves and done a monstrous deed
For which vengeance later would surely be exacted;

And Night bore hateful Doom and black Fate
And Death, and Sleep and the brood of Dreams.
And sleeping with no one, the ebony goddess
Night gave birth to Blame and agonizing Grief,
And to the Hesperides who guard the golden apples
And the fruit-bearing trees beyond glorious Ocean.
And she generated the Destinies and the merciless,
Avenging Fates, Clotho, Lachesis, and Atropos,
Who give mortals at birth good and evil to have,
And prosecute transgressions of mortals and gods.
These goddesses never let up their dread anger
Until the sinner has paid a severe penalty.
And deadly Night bore Nemesis too, more misery
For mortals; and after her, Deception and Friendship
And ruinous Old Age, and hard-hearted Eris.

And hateful Eris bore agonizing Toil,
Forgetfulness, Famine, and tearful Pains,
Battles and Fights, Murders and Manslaughters,
Quarrels, Lying Words and Words Disputatious,
Lawlessness and Recklessness, who share one nature,
And Oath, who most troubles men upon Earth
When anyone willfully swears a false oath.

And Pontos, the Sea, begot his eldest, Nereus,
True and no liar. And they call him Old Man
Because he is unerring and mild, remembers.

THE BIRTH OF THE OLYMPIANS

Later, Kronos forced himself upon Rheia,
And she gave birth to a splendid brood:

Hestia and Demeter and gold-sandalled Hera,
Strong, pitiless Hades, the underworld lord,
The booming Earth-shaker, Poseidon, and finally
Zeus, a wise god, our Father in heaven
Under whose thunder the wide world trembles.

And Kronos swallowed them all down as soon as each
Issued from Rheia's holy womb onto her knees,
With the intent that only he among the proud Ouranians
Should hold the title of King among the Immortals.
For he had learned from Earth and starry Heaven
That it was fated for him, powerful though he was,
To be overthrown by his child, through the scheming of Zeus.
Well, Kronos wasn't blind. He kept a sharp watch
And swallowed his children.
Rheia's grief was unbearable.
When she was about to give birth to Zeus our Father
She petitioned her parents, Earth and starry Heaven,
To put together some plan so that the birth of her child
Might go unnoticed, and she would make devious Kronos
Pay the Avengers of her father and children.
They listened to their daughter and were moved by her words,
And the two of them told her all that was fated
For Kronos the King and his stout-hearted son.
They sent her to Lyktos, to the rich land of Crete,
When she was ready to bear the youngest of her sons,
Mighty Zeus. Vast Earth received him when he was born
To be nursed and brought up in the wide land of Crete.
She came first to Lyktos, travelling quickly by night,
And took the baby in her hands and hid him in a cave,
An eerie hollow in the woods of dark Mount Aigaion.
Then she wrapped up a great stone in swaddling clothes
And gave it to Kronos, Ouranos' son, the great lord and king
Of the earlier gods. He took it in his hands and rammed it
Down into his belly, the poor fool! He had no idea
That a stone had been substituted for his son, who,
Unscathed and content as a babe, would soon wrest
His honors from him by main force and rule the Immortals.
It wasn't long before the young lord was flexing
His glorious muscles. The seasons followed each other,
And great devious Kronos, gulled by Earth's

Clever suggestions, vomited up his offspring,
[Overcome by the wiles and power of his son]
The stone first, which he'd swallowed last.
Zeus took the stone and set it in the ground at Pytho
Under Parnassos' hollows, a sign and wonder for men to come.
And he freed his uncles, other sons of Ouranos
Whom their father in a fit of idiocy had bound.
They remembered his charity and in gratitude
Gave him thunder and the flashing thunderbolt
And lightning, which enormous Earth had hidden before.
Trusting in these he rules mortals and Immortals.

PROMETHEUS

That happened when the gods and mortal men were negotiating
At Mekone. Prometheus cheerfully butchered a great ox
And served it up, trying to befuddle Zeus' wits.
For Zeus he set out flesh and innards rich with fat
Laid out on oxhide and covered with its pauch.
But for others he set out the animal's white bones
Artfully dressed out and covered with shining fat.
And then the Father of gods and men said to him?
"Son of Iapetos, my celebrated lord,
How unevenly you have divided the portions."

Thus Zeus, sneering, with imperishable wisdom.
And Prometheus, whose mind was devious,
Smiled softly and remembered his trickery:

"Zeus most glorious, greatest of the everlasting gods,
Choose whichever of these your heart desires."

This was Prometheus' trick. But Zeus, eternally wise,
Recognized the fraud and began to rumble in his heart
Trouble for mortals, and it would be fulfilled.
With both his hands he picked up the gleaming fat.
Anger seethed in his lungs and bile rose to his heart
When he saw the ox's white bones artfully tricked out.
And that is why the tribes of men on earth
Burn white bones to the immortals upon smoking alters.
But cloudherding Zeus was terribly put out, and said:

"Iapetos' boy, if you're not the smartest of them all.

So you still haven't forgotten your tricks, have you?"

Thus Zeus, angry, whose wisdom never wears out.
From then on he always remembered this trick
And wouldn't give the power of weariless fire
To the ashwood mortals who live on the earth.
But that fine son of Iapetos outwitted him
And stole the far-seen gleam of weariless fire
...and so bit deeply the heart
Of Zeus, the high lord of thunder, who was angry
When he saw the distant gleam of fire among men,
And straight off he gave them trouble to pay for the fire.

PANDORA

The famous Lame God plastered up some clay
To look like a shy virgin, just like Zeus wanted,
And Athena, the Owl-Eyed Goddess,
Got her all dressed up in silvery clothes
And with her hands draped a veil from her head,
An intricate thing, wonderful to look at.
And Pallas Athena circled her head
With a wreath of luscious springtime flowers
And crowned her with a golden tiara
That the famous Lame God had made himself,
Shaped it by hand to please father Zeus,
Intricately designed and a wonder to look at.
Sea monsters and other fabulous beasts
Crowded the surface, and it sighed with beauty,
And you could almost hear the animals' voices.

He made this lovely evil to balance the good,
Then led her off to the other gods and men
Gorgeous in the finery of the owl-eyed daughter
Sired in power. And they were stunned,
Immortal gods and mortal men, when they saw
The sheer deception, irresistible to men.
From her is the race of female women,
The deadly race and population of women,
A great infestation among mortal men,
At home with Wealth but not with Poverty.
It's the same as with bees in their overhung hives
Feeding the drones, evil conspirators.

The bees work every day until the sun goes down,
Busy all day long making pale honeycombs,
While the drones stay inside, in the hollow hives,
Stuffing their stomachs with the work of others.
That's just how Zeus, the high lord of thunder,
Made women as a curse for mortal men,
Evil conspirators. And he added another evil
To offset the good. Whoever escapes marriage
And women's harm, comes to deadly old age
Without any son to support him. He has no lack
While he lives, but when he dies distant relatives
Divide up his estate. Then again, whoever marries
As fated, and gets a good wife, compatible,
Has a life that is balanced between evil and good,
A constant struggle. But if he marries the abusive kind,
He lives with pain in his heart all down the line,
Pain in spirit and mind, incurable evil.
There's no way to get around the mind of Zeus.
Not even Prometheus, that fine son of Iapetos
Escaped his heavy anger. He knows many things,
But he is caught in the crimp of ineluctable bonds.

6 Fragments: 31, 94, 130, 147

Sappho

31

He seems to me equal to gods that man
whoever he is who opposite you
sits and listens close
    to your sweet speaking

and lovely laughing—oh it
puts the heart in my chest on wings
for when I look at you, even a moment, no
  speaking
    is left in me
no: tongue breaks and thin
fire is racing under skin
and in eyes no sight and drumming
    fills ears

and cold sweat holds me and shaking
grips me all, greener than grass
I am and dead—or almost
    I seem to me.
But all is to be dared, because even
a person of
    poverty

94

I simply want to be dead.
Weeping she left me

with many tears and said this:
Oh how badly things have turned
  out for us.
Sappho, I swear, against my will I
  leave you.

And I answered her:
Rejoice, go and
remember me. For you know
  how we cherished
    you.

But if not, I want
to remind you
    ]and beautiful times we had.

For many crowns of violets
and roses
    ]at my side you put on
and many woven garlands
made of flowers

around your soft throat.

And with sweet oil
costly
you anointed yourself

and on a soft bed
delicate
you would let loose your longing

and neither any[ ]nor any
holy place nor
was there from which we were absent

no grove [ ]no dance
]no sound
[

130

Eros the melter of limbs (now again) stirs me—
sweetbitter unmanageable creature who steals in

147

someone will remember us
I say
even in another time

7 *Symposium*: The Speech of Aristophanes[1]

Plato

First you must learn what Human Nature was in the beginning and what has happened to it since, because long ago our nature was not what it is now, but very different. There were three kinds of human beings, that's my first point—not two as there are now, male and female. In addition to these, there was a third, a combination of those two; its name survives, though the kind itself has vanished. At that time, you see, the word "androgynous" really meant something: a form made up of male and female elements, though now there's nothing but the word, and that's used as an insult. My second point is that the shape of each human being was completely round, with back and sides in a circle; they had four hands each, as many legs as hands, and two faces, exactly alike, on a rounded neck. Between the two faces, which were on opposite sides, was one head with four ears. There were two sets of sexual organs, and everything else was the way you'd imagine it from what I've told you. They walked upright, as we do now, whatever direction they wanted. And whenever they set out to run fast, they thrust out all their eight limbs, the ones they had then, and spun rapidly, the way gymnasts do cartwheels, by bringing their legs around straight.

Now here is why there were three kinds, and why they were as I described them: The male kind was originally an offspring of the sun, the female of the earth, and the one that combined both genders was an offspring of the moon, because the moon shares in both. They were spherical, and so was their motion, because they were like their parents in the sky.

In strength and power, therefore, they were terrible, and they had great ambitions. They made an attempt on the gods, and Homer's story about Ephialtes and Otos was originally about them: how they tried to make an ascent to heaven so as to attack the gods.[2] Then Zeus and the other gods met in council to discuss what to do, and they were sore perplexed. They couldn't wipe out the human race with thunderbolts and kill them all off, as they had the giants, because that would wipe out the worship they receive, along with the sacrifices we humans give them. On the other hand, they couldn't let them run riot. At last, after great effort, Zeus had an idea.

"I think I have a plan," he said, "that would allow human beings to exist and stop their misbehaving: they will give up being wicked when they lose their strength. So I shall now cut each of them in two. At one stroke they will lose their strength and also become more profitable to us, owing to the increase in their number. They shall walk upright on two legs. But if I find they still run riot and do not keep the peace," he said, "I will cut them in two again, and they'll have to make their way on one leg, hopping."

So saying, he cut those human beings in two, the way people cut sorb-apples before they dry them or the way they cut eggs with hairs. As he cut each one, he commanded Apollo to turn its face and half its neck towards the wound, so that each person would see that he'd been cut and keep better order. Then Zeus commanded Apollo to heal the rest of the wound, and Apollo did turn the face around, and he drew skin from all sides over what is now called the stomach, and there he made one mouth, as in a pouch with a drawstring, and fastened it at the center of the stomach. This is now called the navel. Then he smoothed out the other wrinkles, of which there were many, and he shaped the breasts, using some such tool as shoemakers have for smoothing wrinkles out of leather on the form. But he left a few wrinkles around the stomach and the navel, to be a reminder of what happened long ago.

Now, since their natural form had been cut in two, each one longed for its own other half, and so they would throw their arms about each other, weaving themselves together, wanting to grow together. In that condition they would die from hunger and general idleness, because they would not do anything apart from each other. Whenever one of the halves died and one was left, the one that was left still sought another and wove itself together with that. Sometimes the half he met came from a woman, as we'd call her now, sometimes it came from a man; either way, they kept on dying.

Then, however, Zeus took pity on them, and came up with another plan: he moved their genitals around to the front! Before then, you see, they used to have their genitals outside, like their faces, and they cast seed and made children, not in one another, but in the ground, like cicadas. So Zeus brought about this relocation of genitals, and in doing so he invented interior reproduction, *by* the man *in* the woman. The purpose of this was so that, when a man embraced a woman, he would cast his seed and they would have children; but when male embraced male, they would at least have the satisfaction of intercourse, after which they could stop embracing, return to their jobs, and look after their other needs in life. This, then, is the source of our desire to love each other. Love is born into every human being; it calls back the halves of our original nature together; it tries to make one out of two and heal the wound of human nature.

Each of us, then, is a "matching half" of a human whole, because each was sliced like a flatfish, two out of one, and each of us is always seeking the half that matches him. That's why a man who is split from the double sort (which used to be called "androgynous") runs after women. Many lecherous men have come from this class, and so do the lecherous women who run after men. Women who are split from a woman, however, pay no attention at all to men; they are oriented more towards women, and lesbians come from this class. People who are split from a male are male-oriented. While they are boys, because they are chips off the male block, they love men and enjoy lying with men and being embraced by men; those are the best of boys and lads, because they are the most manly in their nature. Of course, some say such boys are shameless, but they're lying. It's not because they have no shame that such boys do this, you see, but because they are bold and brave and masculine, and they tend to cherish what is like themselves. Do you want me to prove it? Look, these are the only kind of boys who grow up to be politicians. When they're grown men, they are lovers of young men, and they naturally pay no attention to marriage or to making babies, except insofar as they are required by local custom. They, however, are quite satisfied to live their lives with one another unmarried. In every way, then, this sort of man grows up as a lover of young men and a lover of Love, always rejoicing in his own kind.

And so, when a person meets the half that is his very own, whatever his orientation, whether it's to young men or not, then something wonderful happens: the two are struck from their senses by love, by a sense of belonging to one another, and by desire, and they don't want to be separated from one another, not even for a moment.

These are the people who finish out their lives together and still cannot say what it is they want from one another. No one would think it is the intimacy of sex—that mere sex is the reason each lover takes so great and deep a joy in being with the other. It's obvious that the soul of every lover longs for something else; his soul cannot say what it is, but like an oracle it has a sense of what it wants, and like an oracle it hides behind a riddle. Suppose two lovers are lying together and Hephaestus[3] stands over them with his mending tools, asking, "What is it you human beings really want from each other?" And suppose they're perplexed, and he asks them again: "Is this your heart's desire, then—for the two of you to become parts of the same whole, as near as can be, and never to separate, day or night? Because if that's your desire, I'd like to weld you together and join you into something that is naturally whole, so that the two of you are made into one. Then the two of you would share one life, as long as you lived, because you would be one being, and by the same token, when you died, you would be one and not two in Hades, having died a single death. Look at your love, and see if this is what you desire: wouldn't this be all the good fortune you could want?"

Surely you can see that no one who received such an offer would turn it down; no one would find anything else that he wanted. Instead, everyone would think he'd found out at last what he had always wanted: to come together and melt together with the one he loves, so that one person emerged from two. Why should this be so? It's because, as I said, we used to be complete wholes in our original nature, and now "Love" is the name for our pursuit of wholeness, for our desire to be complete.

Long ago we were united, as I said; but now the god has divided us as punishment for the wrong we did him, just as the Spartans divided the Arcadians.[4] So there's a danger that if we don't keep order before the gods, we'll be split in two again, and then we'll be walking around in the condition of people carved on gravestones in bas-relief, sawn apart between the nostrils, like half dice. We should encourage all men, therefore, to treat the gods with all due reverence, so that we may escape this fate and find wholeness instead. And we will, if Love is our guide and our commander. Let no one work against him. Whoever opposes Love is hateful to the gods, but if we become friends of the god and cease to quarrel with him, then we shall find the young men that are meant for us and win their love, as very few men do nowadays.

Now don't get ideas, Eryximachus, and turn this speech into a comedy. Don't think I'm pointing this at Pausanias and Agathon. Probably, they both do belong to the group that are entirely masculine in nature. But I am speaking about everyone, men and women alike, and I say there's just one way for the human race to flourish: we must bring love to its perfect conclusion, and each of us must win the favors of his very own young man, so that he can recover his original nature. If that is the ideal, then, of course, the nearest approach to it is best in present circumstances, and that is to win the favor of young men who are naturally sympathetic to us.[5]

If we are to give due praise to the god who can give us this blessing, then, we must praise Love. Love does the best that can be done for the time being: he draws us towards what belongs to us. But for the future, Love promises the greatest hope of all: if we treat the gods with due reverence, he will restore to us our original nature, and by healing us, he will make us blessed and happy.

"That," he said, "is my speech about Love, Eryximachus. It is rather different from yours. As I begged you earlier, don't make a comedy of it. I'd prefer to hear what all the others will say-or, rather, what each of them will say, since Agathon and Socrates are the only ones left."

"I found your speech delightful," said Eryximachus, "so I'll do as you say. Really, we've had such a rich feast of speeches on Love, that if I couldn't vouch for the fact that Socrates and Agathon are masters of the art of love, I'd be afraid that they'd have nothing left to say. But as it is, I have no fears on this score."

Then Socrates said, "That's because *you* did beautifully in the contest, Eryximachus. But if you ever get in my position, or rather the position I'll be in after Agathon's spoken so well, then you'll really be afraid. You'll be at your wit's end, as I am now."

"You're trying to bewitch me, Socrates," said Agathon, "by making me think the audience expects great things of my speech, so I'll get flustered."

"Agathon!" said Socrates, "How forgetful do you think I am? I saw how brave and dignified you were when you walked right up to the theater platform along with the actors and looked straight out at that enormous audience. You were about to put your own writing on display, and you weren't the least bit panicked. After seeing that, how could I expect you to be flustered by us, when we are so few?"

"Why, Socrates," said Agathon. "You must think I have nothing but theater audiences on my mind! So you suppose I don't realize that, if you're intelligent, you find a few sensible men much more frightening than a senseless crowd?"

"No," he said, "It wouldn't be very handsome of me to think you crude in any way, Agathon. I'm sure that if you ever run into people you consider wise, you'll pay more attention to them than to ordinary people. But you can't suppose we're in that class; we were at the theater too, you know, part of the ordinary crowd. Still, if you did run into any wise men, other than yourself, you'd certainly be ashamed at the thought of doing anything ugly in front of them. Is that what you mean?"

"That's true," he said.

"On the other hand, you wouldn't be ashamed to do some thing ugly in front of ordinary people. Is that it?"

At that point Phaedrus interrupted: "Agathon, my friend, if you answer Socrates, he'll no longer care whether we get anywhere with what we're doing here, so long as he has a partner for discussion. Especially if he's handsome. Now, like you, I enjoy listening to Socrates in discussion, but it is my duty to see to the praising of Love and to exact a speech from every one of this group. When each of you two has made his offering to the god, then you can have your discussion."

"You're doing a beautiful job, Phaedrus," said Agathon. "There's nothing to keep me from giving my speech. Socrates will have many opportunities for discussion later."

End notes

1 Aristophanes (ca. 450-ca. 385 B.C.) was the famous writer of comedy who satirized Socrates unmercifully in *The Clouds*, on which see Socrates' reaction in the *Apology* at 18D. Here in the *Symposium*, somewhat surprisingly, Plato shows no ill will towards Aristophanes, but supplies him with a masterpiece, an inventive speech that is comic and seriously moving at the same time.

2 *Iliad* v. 385, *Odyssey* xii.308.

3 Hephaestus in Greek mythology is the craftsman god.

4 Arcadia included the city of Mantinea, which opposed Sparta, and was rewarded for this by having its population divided and dispersed in 385 B.C. See Xenophon, *Hellenica* v.2.5-7.

5 Aristophanes began at 193C3 to speak to all men and women, but at C4 and C7 he plainly reverts to the idiom of homosexual love that suits his audience.

8 *Symposium*: The Speech of Socrates

Plato

Now I'll let you go. I shall try to go through for you the speech about Love I once heard from a woman of Maninea, Diotima-a woman who was wise about many things besides this: once she even put off the plague for ten years by telling Athenians what sacrifices to make. She is the one who taught me the art of love, and I shall go through her speech as best I can on my own, using what Agathon and I have agreed to as a basis.

Following your lead, Agathon, one should first describe who love is and what he is like, and afterwards describe his works...

I think it will be easiest for me to proceed the way Diotima did and tell you how she questioned me. You see, I had told her almost the same things Agathon told me just now: that love is a great god and the he belongs to beautiful things. And she used the very same arguments against me that I used against Agathon; she showed how, according to my very own speech, Love is neither beautiful nor good.

So I said, "What do you mean, Diotima? Is love ugly, then, and bad?"

But she said, "Watch your tongue! Do you really think that, if a thing is not beautiful, it has to be ugly?"

"I certainly do."

"And if a thing's not wise, it's ignorant? Or haven't you found out yet that there's something in between wisdom and ignorance?"

"What's that?"

It's judging things correctly without being able to give a reason. Surely you see that this is not the same as knowing—for how could knowledge be unreasoning? And it's not ignorance either—for how could what hits the truth be ignorance? Correct judgement, of course, has this character: it is *in between* understanding and ignorance."

"True," said I, "as you say."

"Then don't force whatever is not beautiful to be ugly, or whatever is not good to be bad. It's the same with love: when you agree he is neither good nor beautiful, you need not think he is ugly and bad; he could be something in between," she said.

"Yet everyone agrees he's a great god," I said.

"Only those who don't know?" she said. "Is that how you mean 'everyone?' Or do you include those who do know?"

"Oh, everyone together."

And she laughed. "Socrates, how could those who say that he's not a god at all agree that he's a great god?"

"Who says that?" I asked.

"You, for one," said she. "Tell me, wouldn't you say that all gods are beautiful and happy? Surely you'd never say a god is not beautiful or happy?"

"Zeus! Not I," I said.

"Well, by calling anyone 'happy,' don't you mean they possess good and beautiful things?"

"Certainly."

"What about love? You agree he needs good and beautiful things, and that's why he desires them—because he needs them."

"I certainly did."

"Then how could he be a god if he has no share in good and beautiful things?"

"There's no way he could, apparently."

"Now do you see? You don't believe love is god either!"

"Then, what could love be?" I asked. "A mortal?"

"Certainly not."

"Then, what is he?"

"He's like what we mentioned before," she said. "He is in between mortal and immortal."

"What do you mean, Diotima?"

"He's a great spirit, Socrates. Everything spiritual, you see, is in between god and mortal."

"What is their function?" I asked.

"They are messengers who shuttle back and forth between the two, conveying prayer and sacrifice from men to gods, while to men they bring commands from the gods and gifts in return for sacrifices. Being in the middle of the two, they round out the whole and bind fast the all to all. Through them all divination passes, through them the art of priests in sacrifice and ritual, in enchantment, prophecy, and sorcery. Gods do not mix with men; they mingle and converse with us through spirits instead, whether we are awake or sleep. He who is wise in any of these ways is a man of the spirits, but he who is wise in any other way, in a profession or any manual work, is merely a mechanic. These spirits are many and various, then, and one of them is love."

"Who are his father and mother?" I asked.

"That's rather a long story," she said. "I'll tell it to you, all the same."

• • •

When Aphrodite was born, the gods held a celebration. Poros, the son of Metis, was there among them. When they had feasted, Penia came begging, as poverty does when there's a party, and stayed by the gates. Now Poros got drunk on nectar (there was no wine yet, you see) and, feeling drowsy, went into the garden of Zeus, where he fell asleep. Then Penia schemed up a plan to relieve her lack of resources: she would get a child from Poros. So she lay beside him and got pregnant with love. That is why love was born to follow Aphrodite and serve her: because he was conceived on the day of her birth. And that's why he is also by nature a lover of beauty, because Aphrodite herself is especially beautiful.

"As the son of Poros and Penia, his lot in life is set to be like theirs. In the first place, he is always poor, and he's far from being delicate and beautiful (as ordinary people think he is); instead, he is tough and shriveled and shoeless and homeless, always lying on the dirt without a bed, sleeping at people's doorsteps and in roadsides under the sky, having his mother's nature, always living with need. But on his father's side he is a schemer after the beautiful and the good; he is brave, impetuous, and intense, an awesome hunter, always weaving snares, resourceful in his pursuit of intelligence, a lover of wisdom through all his life, a genius with enchantments, potions, and clever pleadings.

"He is by nature neither immortal no mortal. But now he springs to life when he gets his way; now he dies—all in the very same day. Because he is his father's son, however, he keeps coming back to life, but then anything he finds his way to always slips away, and for this reason love is never completely without resources, nor is he ever rich.

"He is in between wisdom and ignorance as well. In fact, you see, none of the gods loves wisdom or wants to become wise-for they are wise-and no one else who is wise already loves wisdom; on the other hand, no one who is ignorant will love wisdom either or want to become wise. For what's especially difficult about being ignorant is that you are content with yourself, even though you're neither beautiful and good nor intelligent. If you don't think you need anything, of course you won't want what you don't think you need."

"In that case, Diotima, who *are* the people who love wisdom, if they are neither wise nor ignorant?"

"That's obvious," she said. "A child could tell you. Those who love wisdom fall in between those two extremes. And love is one of them, because he is in love with what is beautiful, and wisdom is extremely beautiful. It follows that love *must* be a lover of wisdom and, as such, is in between being wise and being ignorant. This, too, comes to him from his parentage, from a father who is wise and resourceful and a mother who is not wise and lacks resource.

"My dear Socrates, that, then is the nature of the spirit called love. Considering what you thought about love, it's no surprise that you were led into thinking of love as you did. On the basis of what you say, I conclude that you thought love was *being loved*, rather than *being a lover*. I think that's why love struck you as beautiful in every way: because it is what is really beautiful and graceful that deserves to be loved, and this is perfect and highly blessed; but being a lover takes a different form, which I have just described."

So I said, "All right then, my friend. What you say about love is beautiful, but if you're right, what use is love to human beings?"

"I'll try to teach you that, Socrates, after I finish this. So far I've been explaining the character and the parentage of love. Now, according to you, he is love for beautiful things. But suppose someone asks us, 'Socrates and Diotima, what is the point of loving beautiful things?'

"It's clearer this way: 'The lover of beautiful things has a desire; what does he desire?'"

"That they become his own," I said.

"But that answer calls for still another question, that is, 'What will this man have, when the beautiful things he wants have become his own,"

I said there was no way I could give a ready answer to that question.

Then she said, "Suppose someone changes the question, putting 'good' in place of 'beautiful,' and asks you this: 'Tell me, Socrates, a lover of good things has a desire; what does he desire?'"

"That they become his own," I said.

"And what will he have, when the good things he wants have become his own?"

"This time it's easier to come up with the answer," I said. "He'll have happiness."

"That's what makes happy people happy, isn't it-possessing good things. There's no need to ask further, 'What's the point of wanting happiness?' The answer you gave seems to be final."

"True," I said.

"Now this desire for happiness, this kind of love-do you think it is common to all human beings and that everyone wants to have good things forever and ever? What would you say?"

"Just that," I said. "It is common to all."

"Then, Socrates, why don't we say that everyone is in love," she asked, "since everyone always loves the same things? Instead we say some people are in love and others are not; why is that?"

"I wonder about that myself," I said.

"It's nothing to wonder about," she said. "It's because we divide out a special kind of love, and we refer to it by the word that means the whole-'love'; and for the other kind of love we use other words."

"What do you mean?" I asked.

"Well you know, for example, that 'poetry' has a very wide range, when it is used to mean 'creativity'. After all, everything that is responsible for creating something out of nothing is a kind of poetry; and so all the creations of every craft and profession are themselves a kind of poetry, and everyone who practices a craft is a poet."

"True."

"Nevertheless," she said, "as you also know, these craftsmen are not called poets. We have other words for them, and out of the whole of poetry we have marked off one part, the part the Muses give us with melody and rhythm, and we refer to this by the word that means the whole. For this alone is called 'poetry,' and those who practice this part of poetry are called poets."

"True."

"That's also how it is with love. The main point is this: every desire for good things or for happiness is 'the supreme and treacherous love' in everyone. But those who pursue this along any of its many other ways-through making money, or through the love of sports, or through philosophy-we don't say that *these* people are in love, and don't call them lovers. It's only when people are devoted exclusively to one special kind of love that we use these words that really belong to the whole of it: 'love' and 'in love' and 'lovers.'"

"I am beginning to see your point," I said.

"Now there is a certain story," she said, 'according to which lovers are those people who seek their other halves. But according to my story, a lover does not seek the half or the whole, unless, my friend, it turns out to be good as well. I say this because people are even willing to cut off their own arms and legs if they think they are diseased. I don't think an individual takes joy in what belongs to him personally unless by 'belonging to me' he means 'good' and by 'belonging to another' he means 'bad.' That's because what everyone loves is really nothing other than the good. Do you disagree?"

"Zeus! Not I," I said.

"Now, then," she said. "Can we simply say that people love the good?"

"Yes," I said.

"But shouldn't we add that, in loving it, they want the good to be theirs?"

"We should."

"And not only that," she said. "They want the good to be theirs forever, don't they?"

"We should add that too."

"In a word, then, love is wanting to possess the good forever."

"That's very true," I said.

"This, then, is the object of love," she said. "In view of that, how do people pursue it if they are truly in love? What do they do with the eagerness and zeal we call love? What is the real purpose of love? Can you say?"

"If I could," I said, "I wouldn't be your student, filled with admiration for your wisdom, and trying to learn these very things."

"Well, I'll tell you," she said. "It is giving birth in beauty, whether in body or in soul."

"It would take divination to figure out what you mean. I can't."

"Well, I'll tell you more clearly," she said. "All of us are pregnant, Socrates, both in body and in soul, and, as soon as we come to a certain age, we naturally desire to give birth. Now no one can possibly give birth in anything ugly; only in something beautiful. That's because when a man and a woman come together in order to give birth, this is a godly affair. Pregnancy, reproduction-this is an immortal thing for a mortal animal to do, and it cannot occur in anything that is out of harmony, but ugliness is out of harmony with all that is godly. Beauty, however, is in harmony with the devine. Therefore the goddess who presides at childbirth-she's called Moira or Eileithuia-is really beauty. That's why, whenever pregnant animals or persons draw near to beauty, they become gently and joyfully disposed and give birth and reproduce; but near ugliness they are foulfaced and draw back in pain; they turn away and shrink back and do not reproduce, and because they hold on to what they carry inside them, the labor is painful. This is the source of the great excitement about beauty that comes to anyone who is pregnant and already teeming with life: beauty releases them from their great pain. You see Socrates," she said, "what love wants is not beauty, as you think it is."

"Well, what is it, then?"

"Reproduction and birth in beauty." "Maybe," I said.

"Certainly,' she said. "Now, why reproduction? It's because reproduction goes on forever; it is what mortals have in place of immortality. A lover must desire immortality along with the good, if what we agreed earlier was right, that love wants to possess the good forever. It follows from our argument that love must desire immortality."

All this she taught me, on those occasions when she spoke on the art of love. And once she asked, "what do you think causes love and desire, Socrates? Don't you see what an awful state a wild animal is in when it wants to reproduce? Footed and winged animals alike, all are plagued by the disease of love. First they are sick for intercourse with each other, then for nurturing their young-for their sake the weakest animals stand ready to do battle against the strongest and even to die for them, and they may be racked with famine in order to feed their young. They would do anything for their sake. Human beings, you'd think, would do this because they understand the reason for it; but what causes wild animals to be in such a state of love? Can you say?

And I said again that I didn't know.

So she said, " how do you think you'll ever master the art of love, if you don't know that?"

"But that's why I came to you, Diotima, as I just said. I knew I needed a teacher. So tell me what causes this, and everything else that belongs to the art of love."

"If you really believe that love by its nature aims at what we have often agreed it does, then don't be surprised at the answer," she said. "For among animals the principle is the same as with us, and mortal nature seeks so far as possible to live forever and be immortal. And this is possible in one way only: by reproduction, because it always leaves behind a new young one in place of the old. Even while each living thing is said to be alive and to be the same-as a person is said to be the same from childhood till he turns into an old man-even then he never consists of the same things, though he is called the same, but he is always being renewed and in other respects passing away, in his hair and flesh and bones and blood and his entire body. And it's not just in his body, but in his soul too, for none of his manners, customs, opinions, desires, pleasures, pains, or fears ever remains the same, but some are coming to be in him while others are passing away. And what is still far stranger than that is that not only does one branch of knowledge come to be in us while another passes away and that we are never the same even in respect of our knowledge, but that each single piece of knowledge has the same fate. For what we call studying exists because knowledge is leaving us, because forgetting is the departure of knowledge, while studying puts back a fresh memory in place of what went away, thereby preserving a piece of knowledge, so that it seems to be the same. And in that way everything mortal is preserved, not, like the divine, by always being the same in every way, but because what is departing and aging leaves behind something new, something such as it had been. By this device, Socrates," she said, "what is mortal shares in immortality, whether it is a body or anything else, while the immortal has another way. So don't be surprised if everything naturally values its own offspring, because it is for the sake of immortality that everything shows this zeal, which is Love."

Yet when I heard her speech I was amazed, and spoke: "Well," said I, "Most wise Diotima, is this really the way it is?"

And in the manner of a perfect sophist she said, "Be sure of it, Socrates. Look, if you will, at how human beings seek honor. You'd be amazed at the irrationality, if you didn't have in mind what I spoke about and if you hadn't pondered the awful state of love they're in, wanting to become famous and 'to lay up glory immortal forever,' and how they're ready to brave any danger for the sake of this, much more than they are for their children; and they are prepared to spend money, suffer through all sorts of ordeals, and even die for the sake of glory. Do you really think that Alcestis would have died for Admetus," she asked, "or that Achilles would have died after Patroclus, or that Kodros would have died so as to preserve the throne for his sons, if they hadn't expected the memory of their virtue-which we still hold in honor-to be immortal? Far from it," she said. "I believe that anyone will do anything for the sake of immortal virtue and the glorious fame that follows; and the better the people, the more they will do, for they are all in love with immortality.

"Now, some people are pregnant in body, and for this reason turn more to women and pursue love in that way, providing themselves through childbirth with immortality and remembrance and happiness, as they think, for all time to come; while others are pregnant in soul-because there surely are those who are even more pregnant in their souls that in their bodies, and these are pregnant with what is fitting for a soul to bear and bring to birth. And what is fitting? Wisdom and the rest of virtue, which all poets beget, as well as all the craftsmen who are said to be creative. But by far the greatest and most beautiful part of wisdom deals with the proper ordering of cities and households, and that is called moderation and justice. When someone has been pregnant with these in his soul from early youth, while he is still a virgin, and, having arrive at the proper age, desires to beget and give birth, he too will certainly go about seeking the beauty in which he would beget; for he will never beget in anything ugly. Since he is pregnant, then, he is much more drawn to bodies that are beautiful than to those that are ugly; and if he also has luck to find a soul that is beautiful and noble and well-formed, he is even more drawn to this combination; such a man makes him instantly teem with ideas and arguments about virtue-the qualities a virtuous man should have and the customary activities in which he should engage; and so he tries to educate him. In my view, you see, when he makes contact with someone beautiful and keeps company with him, he conceives and gives birth to what he has been carrying inside him for ages. And whether they are together or apart, he remembers that beauty. And in common with him he nurtures the newborn; such people, therefore, have much more to share than do the parents of human children, and have a firmer bond of friendship, because the children in whom they have a share are more beautiful and more immortal. Everyone would rather have such children that human ones, and would look up to Homer, Hesiod, and the other good poets with envy and admiration for the offspring they have left behind-offspring, which, because they are immortals themselves, provide their parents with immortal glory and remembrance. For example," she said, "those are the sort of children Lycourgos left behind in Sparta as the saviors of Sparta and virtually all of Greece. Among men in other places everywhere, Greek or barbarian, have brought a host of beautiful deeds into the light and begotten every kind of virtue. Already many shrines have sprung up to honor them for their immortal children, which hasn't happened yet to anyone for human offspring.

"Even you, Socrates, could probably come to be initiated into these rights of love. But as for the purpose of these rights when they are done correctly-that is the final and highest mystery, and I don't know if you are capable of it. I myself will tell you," she said, "and I won't stint any effort. And you must try to follow if you can.

"A lover who goes about this matter correctly must begin in his youth to devote himself to beautiful bodies. First, if the leader leads right, he should love one body and beget beautiful ideas there; then he should realize that the beauty of any one body is brother to the beauty of any other and that if he is to pursue beauty of form he'd be very foolish not to think that the beauty of all bodies is one and the same. When he grasps this, he must become a lover of all beautiful bodies, and he must think that this wild gaping after just one body is a small thing and despise it.

"After this he must think that the beauty of people's souls is more valuable than the beauty of their bodies, so that if someone is decent in his soul, even though he is scarcely blooming in his body, our lover must be content to love and care for him and to seek to give birth to such ideas as will make young men better. The result is that our lover will be forced to gaze at the beauty of activities and laws and to see that all this is akin to itself, with the result that he will think that the beauty of bodies is a thing of no importance. After customs he must move on to various kinds of knowledge. The result is that he will see the beauty of knowledge and be looking mainly not at beauty in a single example-as a servant would who favored the beauty of a little boy or a man or a single custom (being a slave, of course, he's low and small-minded)-but the lover is turned to the great sea of beauty, and, gazing upon this, he gives birth to many gloriously beautiful ideas and theories, in unstinting love of wisdom, until, having grown and been strengthened there, he catches sight of such knowledge, and it is the knowledge of such beauty...

"Try to pay attention to me," she said, "as best you can. You see, the man who has been thus far guided in matters of love, who has beheld beautiful things in the right order and correctly, is coming now to the goal of loving: all of a sudden he will catch sight of something wonderfully beautiful in its nature; that, Socrates, is the reason for all his earlier labors:

"First, it always is and neither comes to be nor passes away, neither waxes nor wanes. Second, it is not beautiful this way and ugly that way, nor beautiful at one time and ugly at another, nor beautiful in relation to one thing and ugly in relation to another; nor is it beautiful here but ugly there, as it would be if it were beautiful for some people and ugly for others. Nor will the beautiful appear to him in the guise of a face or hands or anything else that belongs to the body. It will not appear to him as one idea or one kind of knowledge. It is not anywhere in another thing, as in an animal, or in earth, or in heaven, or in anything else, but itself by itself with itself, it is always one in form; and all the other beautiful things share in that, in such a way that when those others come to be or pass away, this does not become the least bit smaller or greater nor suffer any change. So when some one rises by these stages, through loving boys correctly, and begins to see the beauty, he has almost grasped his goal. This is what it is to go aright, or be lead by another, into the mystery of love: one goes always upwards for the sake of this beauty, starting out from beautiful things and using them like rising stairs: from one body to two and from two to all beautiful bodies, then from beautiful bodies to beautiful customs, and from customs to learning beautiful things, and from these lessons he arrives in the end at this lesson, which is learning of this very beauty, so that in the end he comes to know just what it is to be beautiful.

"And there in life, Socrates, my friend," said the woman from Maninea, "there if anywhere should a person live his life, beholding that beauty. If you once see that, it won't occur to you to measure beauty by gold or clothing or beautiful boys and youths-who, if you see them now, strike you out of your senses, and make you, you and many others, eager to be with the boys and love and look at them forever, if there were any way to do that, forgetting food and drink, everything but looking at them and being with them. But how would it be, in our view," she said, "if someone got to see the beautiful itself, absolute, pure, unmixed, not polluted by human flesh or color or any other great nonsense of morality, but if he could see the divine beauty itself in its one form? Do you think it would be a poor life for a human being to look there and to behold it by that which he ought, and to be with it? Or haven't you remembered," she said, "that in that life alone, when he looks at beauty in the only way that beauty can be seen-only then will it become possible for him to give birth not to images of virtue (because he's in touch with no images), but to true virtue (because he is in touch with true beauty). The love of the gods belongs to anyone who has given birth to true virtue and nourished it, and if any human being could become immortal, it would be he."

This, Phaedrus and the rest of you, was what Diotima told me. I was persuaded. And once persuaded, I try to persuade others too that human nature can find no better workmate for acquiring this than love. That's why I say that every man must honor love, why I honor the rites of love myself and practice them with special diligence, and why I commend them to others. Now and always I praise the power and courage of love so far as I am able. Consider this speech, then, Phaedrus, if you wish, a speech in praise of love. Or if not, call it whatever and however you please to call it.

9 *The Hebrew Bible*: The Song of Solomon

The Song of Solomon

[1]The Song of Songs, which is Solomon's.

[2]O that you[a] would kiss me with the
kisses of your[b] mouth!
For your love is better than wine,
[3]your anointing oils are fragrant,
your name is oil poured out;
therefore the maidens love you.
[4]Draw me after you, let us make
haste.
The king has brought me into his
chambers.
We will exult and rejoice in you;
we will extol your love more than
wine;
rightly do they love you.

[5]I am very dark, but comely,
O daughters of Jerusalem,
like the tents of Kedar,
like the curtains of Solomon.
[6]Do not gaze at me because I am
swarthy,
because the sun has scorched me.
My mother's sons were angry with
me,
they made me keeper of the
vineyards;
but, my own vineyard I have not
kept!
[7]Tell me, you whom my soul loves,
where you pasture your flock,
where you make it lie down at
noon;
for why should I be like one who
wanders[c]
beside the flocks of your
companions?

[8]If you do not know,
O fairest among women,
follow in the tracks of the flock,
and pasture your kids
beside the shepherds' tents.

[9]I compare you, my love,
to a mare of Pharaoh's chariots.
[10]Your cheeks are comely with

ornaments,
your neck with strings of jewels.
11We will make you ornaments of gold,
studded with silver.

12While the king was on his couch,
my nard gave forth its fragrance.
13My beloved is to me a bag of myrrh,
that lies between my breasts.
14My beloved is to me a cluster of henna blossoms
in the vineyards of En-ge'di.
15Behold, you are beautiful, my love;
behold, you are beautiful;
your eyes are doves.
16Behold, you are beautiful, my beloved,
truly lovely.
Our couch is green;
17the beams of our house are cedar,
our rafters[d] are pine.

2I am a rose[e] of Sharon,
a lily of the valleys.

2As a lily among brambles,
so is my love among maidens.

3As an apple tree among the trees of the wood,
so is my beloved among young men.
With great delight I sat in his shadow,
and his fruit was sweet to my taste.
4He brought me to the banqueting house,
and his banner over me was love.
5Sustain me with raisins,
refresh me with apples;
for I am sick with love.
6O that his left hand were under my head,
and that his right hand embraced me!
7I adjure you, O daughters of Jerusalem,
by the gazelles or the hinds of the field,
that you stir not up nor awaken love
until it please.

8The voice of my beloved!
Behold, he comes,
leaping upon the mountains,
bounding over the hills.
9My beloved is like a gazelle,
or a young stag.
Behold, there he stands
behind our wall,
gazing in at the windows,
looking through the lattice.
10My beloved speaks and says to me:
"Arise, my love, my fair one,
and come away;
11for lo, the winter is past,
the rain is over and gone.
12The flowers appear on the earth,
the time of singing has come,
and the voice of the turtledove
is heard in our land.
13The fig tree puts forth its figs,
and the vines are in blossom;
they give forth fragrance.
Arise, my love, my fair one,
and come away.
14O my dove, in the clefts of the rock,
in the covert of the cliff,
let me see your face,
let me hear your voice,
for your voice is sweet,
and your face is comely.
15Catch us the foxes,

the little foxes,
that spoil the vineyards,
for our vineyards are in blossom."
[16]My beloved is mine and I am his,
he pastures his flock among the lilies.
[17]Until the day breathes
and the shadows flee,
turn, my beloved, be like a gazelle,
or a young stag upon rugged[f] mountains.

[3]Upon my bed by night
I sought him whom my soul loves;
I sought him, but found him not;
I called him, but he gave no answer.[g]
[2]"I will rise now and go about the city,
in the streets and in the squares;
I will seek him whom my soul loves."
I sought him, but found him not.
[3]The watchmen found me,
as they went about in the city.
"Have you seen him whom my soul loves?"
[4]Scarcely had I passed them,
when I found him whom my soul loves.
I held him, and would not let him go
until I had brought him into my mother's house,
and into the chamber of her that conceived me.
[5]I adjure you, O daughters of Jerusalem,
by the gazelles or the hinds of the field,
that you stir not up nor awaken love
until it please.

[6]What is that coming up from the wilderness,
like a column of smoke,
perfumed with myrrh and frankincense,
with all the fragrant powders of the merchant?
[7]Behold, it is the litter of Solomon!
About it are sixty mighty men
of the mighty men of Israel,
[8]all girt with swords
and expert in war,
each with his sword at his thigh,
against alarms by night.
[9]King Solomon made himself a palanquin
from the wood of Lebanon.
[10]He made its posts of silver,
its back of gold, its seat of purple;
it was lovingly wrought within[h]
by the daughters of Jerusalem.
[11]Go forth, O daughters of Zion,
and behold King Solomon,
with the crown with which his mother crowned him
on the day of his wedding,
on the day of the gladness of his heart.

[4]Behold, you are beautiful, my love,
behold, you are beautiful!
Your eyes are doves
behind your veil.
Your hair is like a flock of goats,
moving down the slopes of Gilead.
[2]Your teeth are like a flock of shorn ewes
that have come up from the washing,
all of which bear twins,
and not one among them is

bereaved.
[3]Your lips are like a scarlet thread,
and your mouth is lovely.
Your cheeks are like halves of a pomegranate
behind your veil.
[4]Your neck is like the tower of David,
built for an arsenal,[i]
whereon hang a thousand bucklers,
all of them shields of warriors.
[5]Your two breasts are like two fawns,
twins of a gazelle,
that feed among the lilies.
[6]Until the day breathes
and the shadows flee,
I will hie me to the mountain of myrrh
and the hill of frankincense.
[7]You are all fair, my love;
there is no flaw in you.
[8]Come with me from Lebanon, my bride;
come with me from Lebanon.
Depart[j] from the peak of Ama'na,
from the peak of Senir and Hermon,
from the dens of lions,
from the mountains of leopards.

[9]You have ravished my heart, my sister, my bride,
you have ravished my heart with a glance of your eyes,
with one jewel of your necklace.
[10]How sweet is your love, my sister, my bride!
how much better is your love than wine,
and the fragrance of your oils than any spice!
[11]Your lips distil nectar, my bride;
honey and milk are under your tongue;
the scent of your garments is like the scent of Lebanon.
[12]A garden locked is my sister, my bride,
a garden locked, a fountain sealed.
[13]Your shoots are an orchard of pomegranates
with all choicest fruits,
henna with nard,
[14]nard and saffron, calamus and cinnamon,
with all trees of frankincense,
myrrh and aloes,
with all chief spices-
[15]a garden fountain, a well of living water,
and flowing streams from Lebanon.

[16]Awake, O north wind,
and come, O south wind!
Blow upon my garden,
let its fragrance be wafted abroad.
Let my beloved come to his garden,
and eat its choicest fruits.

[5]I come to my garden, my sister, my bride,
I gather my myrrh with my spice,
I eat my honeycomb with my honey,
I drink my wine with my milk.

Eat, O friends, and drink:
drink deeply, O lovers!

[2]I slept, but my heart was awake.
Hark! my beloved is knocking.
"Open to me, my sister, my love,
my dove, my perfect one;
for my head is wet with dew,

my locks with the drops of the night."
[3]I had put off my garment,
how could I put it on?
I had bathed my feet,
how could I soil them?
[4]My beloved put his hand to the latch,
and my heart was thrilled within me.
[5]I arose to open to my beloved,
and my hands dripped with myrrh,
my fingers with liquid myrrh,
upon the handles of the bolt.
[6]I opened to my beloved,
but my beloved had turned and gone.
My soul failed me when he spoke.
I sought him, but found him not;
I called him, but he gave no answer.
[7]The watchmen found me,
as they went about in the city;
they beat me, they wounded me,
they took away my mantle,
those watchmen of the walls.
[8]I adjure you, O daughters of Jerusalem,
if you find my beloved,
that you tell him
I am sick with love.

[9]What is your beloved more than another beloved,
O fairest among women?
What is your beloved more than another beloved,
that you thus adjure us?
[10]My beloved is all radiant and ruddy,
distinguished among ten thousand.
[11]His head is the finest gold;
his locks are wavy,
black as a raven.
[12]His eyes are like doves
beside springs of water,
bathed in milk,
fitly set.[k]
[13]His cheeks are like beds of spices,
yielding fragrance.
His lips are lilies,
distilling liquid myrrh.
[14]His arms are rounded gold,
set with jewels.
His body is ivory work,[l]
encrusted with sapphires.[m]
[15]His legs are alabaster columns,
set upon bases of gold.
His appearance is like Lebanon,
choice as the cedars.
[16]His speech is most sweet,
and he is altogether desirable.
This is my beloved and this is my friend,
O daughters of Jerusalem.

[6]Whither has your beloved gone,
O fairest among women?
Whither has your beloved turned,
that we may seek him with you?

[2]My beloved has gone down to his garden,
to the beds of spices,
to pasture his flock in the gardens,
and to gather lilies.
[3]I am my beloved's and my beloved is mine;
he pastures his flock among the lilies.

[4]You are beautiful as Tirzah, my love,
comely as Jerusalem,

terrible as an army with banners.
[5]Turn away your eyes from me,
for they disturb me-
Your hair is like a flock of goats,
moving down the slopes of Gilead.
[6]Your teeth are like a flock of ewes,
that have come up from the
washing,
all of them bear twins,
not one among them is bereaved.
[7]Your cheeks are like halves of a
pomegranate
behind your veil.
[8]There are sixty queens and eighty
concubines,
and maidens without number.
[9]My dove, my perfect one, is only
one,
the darling of her mother,
flawless to her that bore her.
The maidens saw her and called her
happy;
the queens and concubines also,
and they praised her.
[10]"Who is this that looks forth like
the dawn,
fair as the moon, bright as the sun,
terrible as an army with banners?"

[11]I went down to the nut orchard,
to look at the blossoms of the
valley,
to see whether the vines had budded,
whether the pomegranates were
in bloom.
[12]Before I was aware, my fancy set me
in a chariot beside my prince.[n]
[13] [o]Return, return, O Shu'lammite,
return, return, that we may look
upon you.

Why should you look upon the
Shu'lammite,
as upon a dance before two
armies?[p]
[7]How graceful are your feet in
sandals,
O queenly maiden!
Your rounded thighs are like jewels,
the work of a master hand.
[2]Your navel is a rounded bowl
that never lacks mixed wine.
Your belly is a heap of wheat,
encircled with lilies.
[3]Your two breasts are like two fawns,
twins of a gazelle.
[4]Your neck is like an ivory tower.
Your eyes are pools in Heshbon,
by the gate of Bath-rab'bim.
Your nose is like a tower of
Lebanon,
overlooking Damascus.
[5]Your head crowns you like Carmel,
and your flowing locks are like
purple;
a king is held captive in the
tresses.[q]

[6]How fair and pleasant you are,
O loved one, delectable maiden![r]
[7]You are stately[s] as a palm tree,
and your breasts are like its
clusters.
[8]I say I will climb the palm tree
and lay hold of its branches.
Oh, may your breasts be like
clusters of the vine,
and the scent of your breath like
apples,
[9]and your kisses[t] like the best wine
that goes down[u] smoothly,
gliding over lips and teeth.[v]

[10]I am my beloved's,

and his desire is for me.
[11]Come, my beloved,
let us go forth into the fields,
and lodge in the villages;
[12]let us go out early to the vineyards,
and see whether the vines have budded,
whether the grape blossoms have opened
and the pomegranates are in bloom.
There I will give you my love.
[13]The mandrakes give forth fragrance,
and over our doors are all choice fruits,
new as well as old,
which I have laid up for you, O my beloved.
[8]O that you were like a brother to me,
that nursed at my mother's breast!
If I met you outside, I would kiss you,
and none would despise me.
[2]I would lead you and bring you
into the house of my mother,
and into the chamber of her that conceived me.[w]
I would give you spiced wine to drink,
the juice of my pomegranates.
[3]O that his left hand were under my head,
and that his right hand embraced me!
[4]I adjure you, O daughters of Jerusalem,
that you stir not up nor awaken love
until it please.
[5]Who is that coming up from the wilderness,
leaning upon her beloved?
Under the apple tree I awakened you.
There your mother was in travail with you,
there she who bore you was in travail.

[6]Set me as a seal upon your heart,
as a seal upon your arm;
for love is strong as death,
jealousy is cruel as the grave.
Its flashes are flashes of fire,
a most vehement flame.
[7]Many waters cannot quench love,
neither can floods drown it.
If a man offered for love
all the wealth of his house,
it would be utterly scorned.

[8]We have a little sister,
and she has no breasts.
What shall we do for our sister,
on the day when she is spoken for?
[9]If she is a wall,
we will build upon her a battlement of silver;
but if she is a door,
we will enclose her with boards of cedar.
[10]I was a wall,
and my breasts were like towers;
then I was in his eyes
as one who brings[x] peace.

[11]Solomon had a vineyard at Ba'al-ha'mon;
he let out the vineyard to keepers;
each one was to bring for its fruit

a thousand pieces of silver.
[12]My vineyard, my very own, is for myself;
you, O Solomon, may have the thousand,
and the keepers of the fruit two hundred.

[13]O you who dwell in the gardens,
my companions are listening for your voice;
let me hear it.
[14]Make haste, my beloved,
and be like a gazelle
or a young stag
upon the mountains of spices.

End Notes

a Heb *he*
b Heb *his*
c Gk Syr Vg: Heb is *veiled*
d The meaning of the Hebrew word is uncertain
e Heb *crocus*
f The meaning of the Hebrew word is unknown
g Gk: Heb lacks this line
h The meaning of the Hebrew is uncertain
i The meaning of the Hebrew word is uncertain
j Or *Look*
k The meaning of the Hebrew is uncertain
l The meaning of the Hebrew word is uncertain
m Heb *lapis lazuli*
n Cn: The meaning of the Hebrew is uncertain
o Ch 7.1 in Heb
p Or *dance of Mahanaim*
q The meaning of the Hebrew word is uncertain
r Syr: Heb *In delights*
s Heb *This your stature Is*
t Heb *palate*
u Heb *down far my lover*
v Ok Syr Vg: Heb *lips of sleepers*
w Gk Syr: Heb *mother; she* (or *you*) *will teach me*
x Or *finds*

I.I: Title. *Song of songs* is a superlative, like "Holy of holies." *Which is Solomon's*, i.e. either "about Solomon," because he is named in 3.9,11 as the king being wed, or "by Solomon," a late addition based on 1 Kg.4.32.

1.2-4: The maiden longs for her lover. 3: *Your name*, or "you." **4:** *The king*, i.e. the bridegroom. *Has brought me*, the maiden anticipates. *We*, the maiden's companions, a chorus.

1.5-6: The maiden boasts of her rustic beauty, dark as *the tents of Kedar*, sumptuous as *the curtains of Solomon*.

1.7-8: The maiden asks where her lover is, and is answered by the chorus.

1.9-11: The youth praises the bejewelled maiden.

1.12-17: The lovers' dialogue, as they lie together in the open air. *Beloved*, originally an epithet of the fertility god, here simply a term of endearment. In Is.5.1 it is applied to the Lord. **14:** *En-gedi*, a fertile oasis by the Dead Sea.

2.1-4: A second dialogue of the lovers, l: *Rose*, more correctly "crocus." *Sharon*, the rich coastal plain south of Carmel. **4:** *His banner over me*; translation uncertain; perhaps read, "he gazed on me with love."

2.5-7: The maiden's longing for love when the time is ripe.

2.8-14: The lover comes in the spring to summon his bride. 8: *Voice*, or sound of his approach. **11:** *Winter*, the rainy season. **14:** *Clefts* . . . covert, figures of the bride's home.

2.15: Apparently an allusion to what would spoil the luxuriance of love; an isolated fragment.

2.16-17: The maiden delights in her lover's presence. *Until the day breathes*, the breeze of evening.

3.1-5: The maiden dreams of searching for her lover.

3.6-11: A wedding procession of the bridegroom, in the guise of *Solomon*. **10:** Alternative translation of the third line: "Its interior was wrought in leather."

4.1-7: The bridegroom describes the charms of the maiden. 6: *Mountain . . . hill*, her breasts.

4.8-15: The lover bids the maiden accompany him, and praises her lore. 8: These were the mountain dwellings of the Syrian goddess. **12:** *Locked* against any other lover. This verse should follow v. 4.

4.16-5.1: Invitation and response. 5.1: *Drink deeply, O lovers*, or "Be intoxicated with love."

5.2-6.3: The maiden's fruitless search for her lover. A dream like that in 3.1—5, but with a different conclusion.

6.4-10: The groom praises the bride's beauty. 4: *Tirzah*, "the pleasant," once capital of Israel (1 Kg.15.21). *Terrible ... banners*, or "awe-inspiring as banners." 8-9: There may be *sixty queens*, etc., but *my perfect one is the only one*.

6.11-12: The maiden visits the garden.

6.13-7.9: The maiden called on to dance, and her charms arouse the lover's desire. 6.13: *Shulammite*, the meaning of the word is doubtful; it may perhaps mean "bride of Solomon."

7.10-13: The maiden invites her lover to come with her into the fields and promises to give him her love.

8.1-4: The maiden wishes to marry her lover. 1: *Brother*, husband; compare "my sister, my bride" in 4.9.

8.5-7: The lovers return, the maiden imploring her lover to be faithful.

8.8-12: The maiden boasts of her previous chastity.

8.13-14: The lover calls and the maiden answers.

10 *Lyrics Rude & Erotic:* Poems 5, 7, 32, 75

Catullus

5.

Let us live and love,
not listening to old men's talk.
Suns will rise and set
long after our little light
has gone away to darkness.
Kiss me again and again.
Let me kiss you a hundred times,
a thousand more, again a thousand
without rest, losing count, so no
one can speak of us and say
they know the number of our kisses.

7.

Lesbia, you ask me how many
of your kisses are enough
for you to save for me?
As many as there are grains of sand
in the African desert!
As many as the winds that
blow on the shores of Cyrene
where Silphium grows between
King Battus' sacred tomb
and Jupiter Amons Oracle.
As many kisses as there
are stars in the night
that shine on all men's
secrets. After all of these
your Catullus will have
enough kisses for a while,
enough that clever eyes
can not tell the number
and slanderous tongues are
counting so they are mute.

32.

Ipsitilla my sweet
and luscious tart,
ask me to come by at noon.
If you grant me this,
grant me one thing more.
Leave your door unlocked.
Please don't step out,
but prepare for us
a nine course feast of love.
Please send me news at once
for now I lie on my back
and my desire is poking
through the covers bumping
into my breakfast plate.

75.

Lesbia, you are the author of my destruction.
My heart is defeated and weary at the thought of life.
I will wish terrible things for you if you become great,
but I will always love you just the same.

11 *Confessions*: Books II, III

St. Augustine

BOOK II—1

I must now carry my thoughts back to the abominable things I did in those days, the sins of the flesh which defiled my soul. I do this, my God, not because I love those sins, but so that I may love you. For love of your love I shall retrace my wicked ways. The memory is bitter, but it will help me to savour your sweetness, the sweetness that does not deceive but brings real joy and never fails. For love of your love I shall retrieve myself from the havoc of disruption which tore me to pieces when I turned away from you, whom alone I should have sought, and lost myself instead on many a different quest. For as I grew to manhood I was inflamed with desire for a surfeit of hell's pleasures. Foolhardy as I was, I ran wild with lust that was manifold and rank. In your eyes my beauty vanished and I was foul to the core, yet I was pleased with my own condition and anxious to be pleasing in the eyes of men.

2

I cared for nothing but to love and be loved. But my love went beyond the affection of one mind for another, beyond the arc of the bright beam of friendship. Bodily desire, like a morass, and adolescent sex welling up within me exuded mists which clouded over and obscured my heart, so that I could not distinguish the clear light of true love from the murk of lust. Love and lust together seethed within me. In my tender youth they swept me away over the precipice of my body's appetites and plunged me in the whirlpool of sin. More and more I angered you, unawares. For I had been deafened by the clank of my chains, the fetters of the death which was my due to punish the pride in my soul. I strayed still farther from you and you did not restrain me. I was tossed and spilled, floundering in the broiling sea of my fornication, and you said no word. How long it was before I learned that you were my true joy! You were silent then, and I went on my way, farther and farther from you, proud in my distress and restless in fatigue, sowing more and more seeds whose only crop was grief.

Was there no one to lull my distress, to turn the fleeting beauty of these new-found attractions to good purpose and set up a goal for their charms, so that the high tide of my youth might have rolled in upon the shore of marriage? The surge might have been calmed and contented by the procreation of children, which is the purpose of marriage, as your law prescribes, O Lord. By this means you form the offspring of our fallen nature, and with a gentle hand you prune back the thorns that have no place in your paradise. For your almighty power is not far from us, even when we are far from you. Or, again, I might have listened more attentively to your voice from the clouds, saying of those who marry that they *will meet with outward distress, but I leave you your freedom;*[1] *that a man does well to abstain from all commerce with women,*[2] *and that he who is unmarried is concerned with God's claim, asking how he is to please God; whereas the married man is concerned with the world's claim, asking how he is to please his wife.*[3] These were the words to which I should have listened with more care, and if I had made myself a *eunuch for love of the kingdom of heaven,*[4] I should have awaited your embrace with all the greater joy.

But, instead, I was in a ferment of wickedness. I deserted you and allowed myself to be carried away by the sweep of the tide. I broke all your lawful bounds and did not escape your lash. For what man can escape it? You were always present, angry and merciful at once, strewing the pangs of bitterness over all my lawless pleasures to lead me on to look for others unallied with pain. You meant me to find them nowhere but in yourself, O Lord, for you teach us by inflicting pain,[5] you smite so that you may heal,[6] and you kill us so that we may not die away from you. Where was I then and how far was I banished from the bliss of your house in that sixteenth year of my life? This was the age at which the frenzy gripped me and I surrendered myself entirely to lust, which your law forbids but human hearts are not ashamed to sanction. My family made no effort to save me from my fall by marriage. Their only concern was that I should learn how to make a good speech and how to persuade others by my words.

3

In the same year my studies were interrupted. I had already begun to go to the nearby town of Madaura to study literature and the art of public speaking, but I was brought back home while my father, a modest citizen of Thagaste whose determination was greater than his means, saved up the money to send me farther afield to Carthage. I need not tell all this to you, my God, but in your presence I tell it to my own kind, to those other men, however few, who may perhaps pick up this book. And I tell it so that I and all who read my words may realize the depths from which we are to cry to you. Your ears will surely listen to the cry of a penitent heart which lives the life of faith.

No one had anything but praise for my father who, despite his slender resources, was ready to provide his son with all that was needed to enable him to travel so far for the purpose of study. Many of our townsmen, far richer than my father, went to no such trouble for their children's sake. Yet this same father of mine took no trouble at all to see how I was growing in your sight or whether I was chaste or not. He cared only that I should have a fertile tongue, leaving my heart to bear none of your fruits, my God, though you are the only Master, true and good, of its husbandry.

In the meanwhile, during my sixteenth year, the narrow means of my family obliged me to leave school and live idly at home with my parents. The brambles of lust grew high above my head and there was no one to root them out, certainly not my father. One day at the public baths he saw the signs of active virility coming to life in me and this was enough to make him relish the thought of having grandchildren. He was happy to tell my mother about it, for his happiness was due to the intoxication which causes the world to forget you, its Creator, and to love the things you have created instead of loving you, because the world is drunk with the invisible wine of its own perverted, earthbound will. But in my mother's heart you had already begun to build your temple and laid the foundations of your holy dwelling, while my father was still a catechumen and a new one at that. So, in her piety, she became alarmed and apprehensive, and although I had not yet been baptized, she began to dread that I might follow in the crooked path of those who do not keep their eyes on you but turn their backs instead.

How presumptuous it was of me to say that you were silent, my God, when I drifted farther and farther away from you! Can it be true that you said nothing to me at that time? Surely the words which rang in my ears, spoken by your faithful servant, my mother, could have come from none but you? Yet none of them sank into my heart to make me do as you said. I well remember what her wishes were and how she most earnestly warned me not to commit fornication and above all not to seduce any man's wife. It all seemed womanish advice to me and I should have blushed to accept it. Yet the words were yours, though I did not know it. I thought that you were silent and that she was speaking, but all the while you were speaking to me through her, and when I disregarded her, your handmaid, I was disregarding you, though I was both her son and your servant. But I did this unawares and continued headlong on my way. I was so blind to the truth that among my companions I was ashamed to be less dissolute than they were. For I heard them bragging of their depravity, and the greater the sin the more they gloried in it, so that I took pleasure in the same vices not only for the enjoyment of what I did, but also for the applause I won.

Nothing deserves to be despised more than vice; yet I gave in more and more to vice simply in order not to be despised. If I had not sinned enough to rival other sinners, I used to pretend that I had done things I had not done at all, because I was afraid that innocence would be taken for cowardice and chastity for weakness. These were the companions with whom I walked the streets of Babylon. I wallowed in its mire as if it were made of spices and precious ointments, and to fix me all the faster in the very depths of sin the unseen enemy trod me underfoot and enticed me to himself, because I was an easy prey for his seductions. For even my mother, who by now had escaped from the centre of Babylon, though she still loitered in its outskirts, did not act upon what she had heard about me from her husband with the same earnestness as she had advised me about chastity. She saw that I was already infected with a disease that would become dangerous later on, but if the growth of my passions could not be cut back to the quick, she did not think it right to restrict them to the bounds of married love. This was because she was afraid that the bonds of marriage might be a hindrance to my hopes for the future not of course the hope of the life to come, which she reposed in you, but my hopes of success at my studies. Both my parents were unduly eager for me to learn, my father because he gave next to no thought to you and only shallow thought to me, and my mother because she thought that the usual course of study would certainly not hinder me, but would even help me, in my approach to you. To the best of my memory this is how I construe the characters of my parents. Furthermore, I was given a free rein to amuse myself beyond the strict limits of discipline, so that I lost myself in many kinds of evil ways, in all of which a pall of darkness hung between me and the bright light of your truth, my God. What malice proceeded from my pampered heart![7]

4

It is certain, O Lord, that theft is punished by your law, the law that is written in men's hearts and cannot be erased however sinful they are. For no thief can bear that another thief should steal from him, even if he is rich and the other is driven to it by want. Yet I was willing to steal, and steal I did; although I was not compelled by any lack, unless it were the lack of a sense of justice or a distaste for what was right and a greedy love of doing wrong. For of what I stole I already had plenty, and much better at that, and I had no wish to enjoy the things I coveted by stealing, but only to enjoy the theft itself and the sin. There was a pear tree near our vineyard, loaded with fruit that was attractive neither to look at nor to taste. Late one night a band of ruffians, myself included, went off to shake down the fruit and carry it away, for we had continued our games out of doors until well after dark, as was our pernicious habit. We took away an enormous quantity of pears, not to eat them ourselves, but simply to throw them to the pigs. Perhaps we ate some of them, but our real pleasure consisted in doing something that was forbidden.

Look into my heart, O God, the same heart on which you took pity when it was in the depths of the abyss. Let my heart now tell you what prompted me to do wrong for no purpose, and why it was only my own love of mischief that made me do it. The evil in me was foul, but I loved it. I loved my own perdition and my own faults, not the things for which I committed wrong, but the wrong itself. My soul was vicious and broke away from your safe keeping to seek its own destruction, looking for no profit in disgrace but only for disgrace itself.

5

The eye is attracted by beautiful objects, by gold and silver and all such things. There is great pleasure, too, in feeling something agreeable to the touch, and material things have various qualities to please each of the other senses. Again, it is gratifying to be held in esteem by other men and to have the power of giving them orders and gaining the mastery over them. This is also the reason why revenge is sweet. But our ambition to obtain all these things must not lead us astray from you, O Lord, nor must we depart from what your law allows. The life we live on earth has its own attractions as well, because it has a certain beauty of its own in harmony with all the rest of this world's beauty. Friendship among men, too, is a delightful bond, uniting many souls in one. All these things and their like can be occasions of sin because, good though they are, they are of the lowest order of good, and if we are too much tempted by them we abandon those higher and better things, your truth, your law, and you yourself, O Lord our God. For these earthly things, too, can give joy, though not such joy as my God, who made them all, can give, because *honest men will rejoice in the Lord; upright hearts will not boast in vain.*[8]

When there is an inquiry to discover why a crime has been committed, normally no one is satisfied until it has been shown that the motive might have been either the desire of gaining, or the fear of losing, one of those good things which I said were of the lowest order. For such things are attractive and have beauty, although they are paltry trifles in comparison with the worth of God's blessed treasures. A man commits murder and we ask the reason. He did it because he wanted his victim's wife or estates for himself, or so that he might live on the proceeds of robbery, or because he was afraid that the other might defraud him of something, or because he had been wronged and was burning for revenge. Surely no one would believe that he would commit murder for no reason but the sheer delight of killing? Sallust tells us that Catiline was a man of insane ferocity, 'who chose to be cruel and vicious without apparent reason';[9] but we are also told that his purpose was 'not to allow his men to lose heart or waste their skill through lack of practice'.[10] If we ask the reason for this, it is obvious that he meant that once he had made himself master of the government by means of this continual violence, he would obtain honour, power, and wealth and would no longer go in fear of the law because of his crimes or have to face difficulties through lack of funds. So even Catiline did not love crime for crime's sake. He loved something quite different, for the sake of which he committed his crimes.

6

If the crime of theft which I committed that night as a boy of sixteen were a living thing, I could speak to it and ask what it was that, to my shame, I loved in it. I had no beauty because it was a robbery. It is true that the pears which we stole had beauty, because they were created by you, the good God, who are the most beautiful of all beings and the Creator of all things, the supreme Good and my own true Good. But it was not the pears that my unhappy soul desired. I had plenty of my own, better than those, and I only picked them so that I might steal. For no sooner had I picked them than I threw them away, and tasted nothing in them but my own sin, which I relished and enjoyed. If any part of one of those pears passed my lips, it was the sin that gave it flavour.

And now, O Lord my God, now that I ask what pleasure I had in that theft, I find that it had no beauty to attract me. I do not mean beauty of the sort that justice and prudence possess, nor the beauty that is in man's mind and in his memory and in the life that animates him, nor the beauty of the stars in their allotted places or of the earth and sea, teeming with new life born to replace the old as it passes away. It did not even have the shadowy, deceptive beauty which makes vice attractive - pride, for instance, which is a pretence of superiority, imitating yours, for you alone are God, supreme over all; or ambition, which is only a craving for honour and glory, when you alone are to be honoured before all and you alone are glorious for ever. Cruelty is the weapon of the powerful, used to make others fear them: yet no one is to be feared but God alone, from whose power nothing can be snatched away or stolen by any man at any time or place or by any means. The lustful use caresses to win the love they crave for, yet no caress is sweeter than your charity and no love is more rewarding than the love of your truth, which shines in beauty above all else. Inquisitiveness has all the appearance of a thirst for knowledge, yet you have supreme knowledge of all things. Ignorance, too, and stupidity choose to go under the mask of simplicity and innocence, because you are simplicity itself and no innocence is greater than yours. You are innocent even of the harm which overtakes the wicked, for it is the result of their own actions. Sloth poses as the love of peace: yet what certain peace is there besides the Lord? Extravagance masquerades as fullness and abundance: but you are the full, unfailing store of never-dying sweetness. The spendthrift makes a pretence of liberality: but you are the most generous dispenser of all good. The covetous want many possessions for themselves: you possess all. The envious struggle for preferment: but what is to be preferred before you? Anger demands revenge: but what vengeance is as just as yours? Fear shrinks from any sudden, unwonted danger which threatens the things that it loves, for its only care is safety: but to you nothing is strange, nothing unforeseen. No one can part you from the things that you love, and safety is assured nowhere but in you. Grief eats away its heart for the loss of things which it took pleasure in desiring, because it wants to be like you, from whom nothing can be taken away.

So the soul defiles itself with unchaste love when it turns away from you and looks elsewhere for things which it cannot find pure and unsullied except by returning to you. All who desert you and set themselves up against you merely copy you in a perverse way; but by this very act of imitation they only show that you are the Creator of all nature and, consequently, that there is no place whatever where man may hide away from you.

What was it, then, that pleased me in that act of theft? Which of my Lord's powers did I imitate in a perverse and wicked way? Since I had no real power to break his law, was it that I enjoyed at least the pretence of doing so, like a prisoner who creates for himself the illusion of liberty by doing something wrong, when he has no fear of punishment, under a feeble hallucination of power? Here was the slave who ran away from his master and chased a shadow instead! What an abomination! What a parody of life! What abysmal death! Could I enjoy doing wrong for no other reason than that it was wrong?

7

What return shall I make to the Lord[11] for my ability to recall these things with no fear in my soul? I will love you, Lord, and thank you, and praise your name, because you have forgiven me such great sins and such wicked deeds. I acknowledge that it was by your grace and mercy that you melted away my sins like ice. I acknowledge, too, that by your grace I was preserved from whatever sins I did not commit, for there was no knowing what I might have done, since I loved evil even if it served no purpose. I avow that you have forgiven me all, both the sins which I committed of my own accord and those which by your guidance I was spared from committing.

What man who reflects upon his own weakness can dare to claim that his own efforts have made him chaste and free from sin, as though this entitled him to love you the less, on the ground that he had less need of the mercy by which you forgive the sins of the penitent? There are some who have been called by you and because they have listened to your voice they have avoided the sins which I here record and confess for them to read. But let them not deride me for having been cured by the same Doctor who preserved them from sickness, or at least from such grave sickness as mine. Let them love you just as much, or even more, than I do, for they can see that the same healing hand which rid me of the great fever of my sins protects them from falling sick of the same disease.

8

It brought me no happiness, *for what harvest did I reap from acts which now make me blush*,[12] particularly from that act of theft? I loved nothing in it except the thieving, though I cannot truly speak of that as a 'thing' that I could love, and I was only the more miserable because of it. And yet, as I recall my feelings at the time, I am quite sure that I would not have done it on my own. Was it then that I also enjoyed the company of those with whom I committed the crime? If this is so, there was something else I loved besides the act of theft; but I cannot call it 'something else', because companionship, like theft, is not a thing at all.

No one can tell me the truth of it except my God, who enlightens my mind and dispels its shadows. What conclusion am I trying to reach from these questions and this discussion? It is true that if the pears which I stole had been to my taste, and if I had wanted to get them for myself, I might have committed the crime on my own if I had needed to do no more than that to win myself the pleasure. I should have had no need to kindle my glowing desire by rubbing shoulders with a gang of accomplices. But as it was not the fruit that gave me pleasure, I must have got it from the crime itself, from the thrill of having partners in sin.

9

How can I explain my mood? It was certainly a very vile frame of mind and one for which I suffered; but how can I account for it? *Who knows his own frailties?*[13]

We were tickled to laughter by the prank we had played, because no one suspected us of it although the owners were furious. Why was it, then, that I thought it fun not to have been the only culprit? Perhaps it was because we do not easily laugh when we are alone. True enough: but even when a man is all by himself and quite alone, sometimes he cannot help laughing if he thinks or hears or sees something especially funny. All the same, I am quite sure that I would never have done this thing on my own.

My God, I lay all this before you, for it is still alive in my memory. By myself I would not have committed that robbery. It was not the takings that attracted me but the raid itself, and yet to do it by myself would have been no fun and I should not have done it. This was friendship of a most unfriendly sort, bewitching my mind in an inexplicable way. For the sake of a laugh, a little sport, I was glad to do harm and anxious to damage another; and that without thought of profit for myself or retaliation for injuries received! And all because we are ashamed to hold back when others say 'Come on! Let's do it!'

10

Can anyone unravel this twisted tangle of knots? I shudder to look at it or think of such abomination. I long instead for innocence and justice, graceful and splendid in eyes whose sight is undefiled. My longing fills me and yet it cannot cloy. With them is certain peace and life that cannot be disturbed. The man who enters their domain goes to *share the joy of his Lord*.[14] He shall know no fear and shall lack no good. In him that is goodness itself he shall find his own best way of life. But I deserted you, my God. In my youth I wandered away, too far from your sustaining hand, and created of myself a barren waste.

BOOK III—1

I went to Carthage, where I found myself in the midst of a hissing cauldron of lust. I had not yet fallen in love, but I was in love with the idea of it, and this feeling that something was missing made me despise myself for not being more anxious to satisfy the need. I began to look around for some object for my love, since I badly wanted to love something. I had no liking for the safe path without pitfalls, for although my real need was for you, my God, who are the food of the soul, I was not aware of this hunger. I felt no need for the food that does not perish, not because I had had my fill of it, but because the more I was starved of it the less palatable it seemed. Because of this my soul fell sick. It broke out in ulcers and looked about desperately for some material, worldly means of relieving the itch which they caused. But material things, which have no soul, could not be true objects for my love. To love and to have my love returned was my heart's desire, and it would be all the sweeter if I could also enjoy the body of the one who loved me.

So I muddied the stream of friendship with the filth of lewdness and clouded its clear waters with hell's black river of lust. And yet, in spite of this rank depravity, I was vain enough to have ambitions of cutting a fine figure in the world. I also fell in love, which was a snare of my own choosing. My God, my God of mercy, how good you were to me, for you mixed much bitterness in that cup of pleasure! My love was returned and finally shackled me in the bonds of its consummation. In the midst of my joy I was caught up in the coils of trouble, for I was lashed with the cruel, fiery rods of jealousy and suspicion, fear, anger, and quarrels.

End Notes

1 1 Cor. 7: 28.
2 1 Cor. 7: 1.
3 1 Cor. 7: 32, 33.
4 Matt. 19: 12.
5 See Ps. 93: 20 (94: 20).
6 See Deut. 33: 39.
7 See Ps. 72: 7 (73:7).
8 Ps. 63: 11 (64: 10).
9 Sallust, Catilina xvl
10 Sallust, Catitina xvl
11 Ps. 115: 12 (116: 12).
12 Rom. 6: 21.
13 Ps. 18:13 (19:12).
14 Matt. 25:21.

12 *The Holy Qur'an*: Sura 4: Women

A Medinan sura which takes its title from the many references to women throughout the sura (verses 3-4, 127-30). It gives a number of instructions, urging justice to children and orphans, and mentioning inheritance and marriage laws. The early verses of the sura give rulings on property and inheritance, and so does the verse which concludes the sura. The sura also talks of the tensions between the Muslim community in Medina and some of the People of the Book (verses 44, 61), moving into a general discussion of war: it warns the Muslims to be cautious and to defend the weak and helpless (verses 71-6). Another similar theme is the intrigues of the hypocrites (verses 88-91, 138-46).

In the name of God, the Lord of Mercy, the Giver of Mercy

People, be mindful of your Lord, who created you from a single soul, and from it[a] created its mate, and from the pair of them spread countless men and women far and wide; be mindful of God, in whose name you make requests of one another. Beware of severing the ties of kinship:[b] God is always watching over you. Give orphans their property, do not replace [their] good things with bad, and do not consume their property with your own—a serious crime. If you fear that you will not deal fairly with orphan girls,[c] you may marry whichever [other][d] women seem good to you, two, three, or four. If you fear that you cannot be equitable [to them], then marry only one, or your slave(s):[e] that is more likely to make you avoid bias. Give women their dowry as a gift upon marriage, though if they are happy to give up some of it for you, you may enjoy it with clear conscience.

Do not entrust your property to the feeble-minded. God has made it a means of support for you: make provision for them from it, clothe them, and address them kindly. Test orphans until they reach marriageable age; then, if you find they have sound judgement, hand over their property to them. Do not consume it wastefully before they come of age: if the guardian is well off he should abstain from the orphan's property, and if he is poor he should use only what is fair. When you give them their property, call witnesses in; God takes full account of everything you do.

Men shall have a share in what their parents and closest relatives leave, and women shall have a share in what their parents and closest relatives leave, whether the legacy be small or large: this is ordained by God. If other relatives, orphans, or needy people are present at the distribution, give them something too, and speak kindly to them. Let those who would fear for the future of their own helpless children, if they were to die, show the same concern [for orphans]; let them be mindful of God and speak out for justice. Those who consume the property of orphans unjustly are actually swallowing fire into their own bellies: they will burn in the blazing Flame.

Concerning your children, God commands you that a son should have the equivalent share of two daughters. If there are only daughters, two or more should share two-thirds of the inheritance, if one, she should have half. Parents inherit a sixth each if the deceased leaves children; if he leaves no children and his parents are his sole heirs, his mother has a third, unless he has brothers, in which case she has a sixth. [In all cases, the distribution comes] after payment of any bequests or debts. You cannot know which of your parents or your children is closer to you in benefit: this is a law from God, and He is all knowing, all wise. You inherit half of what your wives leave, if they have no children; if they have children, you inherit a quarter. [In all cases, the distribution comes] after payment of any bequests or debts. If you have no children, your wives' share is a quarter; if you have children, your wives get an eighth. [In all cases, the distribution comes] after payment of any bequests or debts. If a man or a woman dies leaving no children or parents,[f] but a single brother or sister, he or she should take one-sixth of the inheritance; if there are more siblings, they share one-third between them. [In all cases, the distribution comes] after payment of any bequests or debts, with no harm done to anyone: this is a commandment from God, and He is all knowing and benign to all. These are the bounds set by God: God will admit those who obey Him and His Messenger to Gardens graced with flowing streams, and there they will stay that is the supreme triumph! But those who disobey God and His Messenger and overstep His limits will be consigned by God to the Fire, and there they will stay—a humiliating torment awaits them!

If any of your women commit a lewd act, call four witnesses from among you, then, if they testify to their guilt, keep the women at home until death comes to them or until God gives them another way out.[g] If two men commit a lewd act, punish them both; if they repent and mend their ways, leave them alone—God is always ready to accept repentance, He is full of mercy. But God only undertakes to accept repentance from those who do evil out of ignorance and soon afterwards repent: these are the ones God will forgive, He is all knowing, all wise. It is not true repentance when people continue to do evil until death confronts them and then say, 'Now I repent,' nor when they die as disbelievers: We have prepared a painful torment for these.

You who believe, it is not lawful for you to inherit women against their will,[h] nor should you treat your wives harshly, hoping to take back some of the dowry you gave them, unless they are guilty of something clearly outrageous. Live with them in accordance with what is fair and kind: if you dislike them, it may well be that you dislike something in which God has put much good. If you wish to replace one wife with another, do not take any of her dowry back, even if you have given her a great amount of gold. How could you take it when this is unjust and a blatant sin? How could you take it when you have lain with each other and they have taken a solemn pledge from you?

Do not marry women that your fathers married—with the exception of what is past—this is indeed a shameful thing to do, loathsome and leading to evil. You are forbidden to take as wives your mothers, daughters, sisters, paternal and maternal aunts, the daughters of brothers and daughters of sisters, your milk-mothers and milk-sisters,[i] your wives' mothers, the stepdaughters in your care—those born of women with whom you have consummated marriage, if you have not consummated the marriage, then you will not be blamed—wives of your begotten sons, two sisters simultaneously—with the exception of what is past: God is most forgiving and merciful—women already married, other than your slaves: God has ordained all this for you. Other women are lawful to you, so long as you seek them in marriage, with gifts from your property, looking for wedlock rather than fornication. If you wish to enjoy women through marriage, give them their dowry—this is obligatory—though if you should choose mutually, after fulfilling this obligation, to do otherwise [with the dowry], you will not be blamed: God is all knowing and all wise.

If any of you does not have the means to marry a believing free woman, then marry a believing slave—God knows best [the depth] of your faith: you are [all] part of the same family.[k] So marry them with their people's consent and their proper dowries. [Make them] married women, not adulteresses or lovers. If they commit adultery when they are married, their punishment will be half that of free women. Only those of you who fear that they will sin should marry slaves; it is better for you to practise self-restraint. God is most forgiving and merciful. He wishes to make His laws clear to you and guide you to the righteous ways of those who went before you. He wishes to turn towards you in mercy—He is all knowing, all wise—He wishes to turn towards you, but those who follow their lusts want you to go far astray. God wishes to lighten your burden; man was created weak.

You who believe, do not wrongfully consume each other's wealth but trade by mutual consent. Do not kill each other, for God is merciful to you. If any of you does these things, out of hostility and injustice, We shall make him suffer Fire: that is easy for God. But if you avoid the great sins you have been forbidden, We shall wipe out your minor misdeeds and let you in through the entrance of honour. Do not covet what God has given to some of you more than others—men have the portion they have earned and women their portion—you should rather ask God for some of His bounty: He has full knowledge of everything. We have appointed heirs for everything that parents and close relatives leave behind, including those to whom you have pledged your hands [in marriage], so give them their share: God is witness to everything.

Husbands should take full care of their wives, with [the bounties] God has given to some more than others and with what they spend out of their own money. Righteous wives, are devout and guard what God would have them guard in their husbands' absence. If you fear high-handedness[m] from your wives, remind them [of the teachings of God], then ignore them when you go to bed, then hit them.[n] If they obey you, you have no right to act against them: God is most high and great. If you [believers] fear that a couple may break up, appoint one arbiter from his family and one from hers. Then, if the couple want to put things right, God will bring about a reconciliation between them: He is all knowing, all aware.

Worship God; join nothing with Him. Be good to your parents, to relatives, to orphans, to the needy, to neighbours near and far, to travellers in need, and to your slaves. God does not like arrogant, boastful people, who are miserly and order other people to be the same, hiding the bounty God has given them. We have prepared a humiliating torment for such ungrateful people. [Nor does He like those] who spend their wealth to show off, who do not believe in Him or the Last Day. Whoever has Satan as his companion has an evil companion! What harm would it do them to believe in God and the Last Day, and give charitably from the sustenance God has given them? God knows them well. He does not wrong anyone by as much as the weight of a speck of dust: He doubles any good deed and gives a tremendous reward of His own. What will they do when We bring a witness from each community, with you [Muhammad] as a witness against these people? On that day, those who disbelieved and disobeyed the Prophet will wish that the earth could swallow them up: they will not be able to hide anything from God.

End notes

A 'From the same essence'. Razi convincingly reached this conclusion based on comparison with many instances when *min anfusikum* is used in the Qur'an.

B Literally 'the womb-relationships', i.e. all those to whom you are related, This expression occurs again in 47: 22.

C In pre-Islamic Arabia, some guardians of orphan girls used to marry them in order to take their property (see 4: 127).

D This is a widely accepted interpretation.

E 'Literally 'what your right hands possess'.

F This is the most generally accepted meaning of the Arabic word *kalala*. There are many others

G Through another regulation, or marriage, or any other way. See also end of 65: 2, which uses nearly identical words.

H In pre-Islamic Arabia, if a man died leaving a widow, her stepson or another man of his family could inherit her

I Islam regards women who breastfeed other people's infants as their 'milk- mothers', not merely 'wet nurses'.

J Slave women were often unclaimed war captives, who would not be in a position to dissolve any previous marriage. An owner was not permitted to touch a slave woman whose husband was with her (Abu Hanifa, in Razi).

K Literally 'you are from one another'.

L The preposition *min* here is taken to have an explanatory rather than a partitive function, which would render the translation 'some of what they have earned'.

M The verb *nashaza* from which *nushuz* is derived means 'to become high', 'to rise'. See also verse 128, where the same word is applied to husbands. It applies to a situation where one partner assumes superiority to the other and behaves accordingly.

N See Abdel Haleem, *Understanding the Quran*, 46-54.

13 “Love Supreme: Gay Nuptials and the Making of Modern Marriage”

Adam Haslett

In December of 1990, Genora Dancel and Ninia Baehr, a lesbian couple from Honolulu, applied for and were denied a license to marry. They decided to file suit against the state for discrimination. The local branch of the A.C.L.U. declined to represent them, and the national gay legal organizations initially kept their distance, considering the issue premature. But the couple persisted, and three years later the Hawaii Supreme Court became the first in the nation to support the right of same-sex couples to wed. Conservative religious groups poured money into the state and eventually helped pass an amendment to its constitution declaring marriage an exclusively heterosexual institution.

More was at stake than the laws of Hawaii. Article IV of the United States Constitution establishes that “Full Faith and Credit shall be given in each State to the public Acts, Records, and judicial Proceedings of every other State.” This is the reason that a couple married in New York can fly to California and still legally be husband and wife when they land. Worried that other states might one day extend marriage rights to same-sex couples, conservatives in Congress introduced the Defense of Marriage Act, which defined marriage as between a man and a woman for the purposes of all federal law, from taxes to Social Security, and released the states from any constitutional obligation to recognize same-sex marriages that might be performed elsewhere. President Clinton, seeking to deny Bob Dole a wedge issue in his 1996 reelection campaign, signed the bill late on a Friday night, after the press corps had gone home.

The issue’s sudden prominence during the past several months stems from a decision of the Supreme Judicial Court of Massachusetts last November to grant same-sex couples full civil marriage. For the first time in this country’s history, a state has sanctioned the marriage of two men or two women; because the Massachusetts constitution can’t be amended any sooner than 2006—the process is under way but by no means certain to succeed—it will do so for at least the next two years. The Bay State thus joins the Netherlands, Belgium, and three Canadian provinces in offering gay couples not only the rights and obligations of marriage but the word itself.

The ensuing national ferment has become a struggle over the meaning and purpose of matrimony. For marriage as we know it today is the product of a particular history—a history that explains both its public character and the private expectations we have for it. The advent of same-sex marriage brings into focus a much larger transformation in how we have come to imagine the institution.

For centuries in Europe, formal marriage was a private contract between landed families, designed to insure that property remained within a particular lineage. In the upper classes, families essentially married other families, forging political alliances and social obligations among relatives and kin. It was during the Reformation, with the emergence of the early Protestant idea of "companionate marriage," that the emotional bond between husband and wife came to be seen as an end in itself. As the social historian Lawrence Stone noted, this was a marked departure from the Catholic ideal of chastity, which considered earthly marriage a more or less unfortunate necessity meant to accommodate human weakness; "It is better to marry than to burn," St. Paul had said, but he made it sound like a close call. So when the Puritans wrote of husbands and wives as mutually respectful and affectionate partners they were moving toward a new understanding of marriage as a kind of spiritual friendship.

It was Milton who took this concept to its logical conclusion. Having married a woman with whom he soon discovered he had nothing in common, he became a staunch advocate of divorce. When the "meet and happy conversation" that is "the chiefest and the noblest end of marriage" ceases, he argued, no authority should have the power to force a man and a woman to remain wed. It was hundreds of years before the law caught up with his notion that irreconcilable differences might be grounds for divorce. But what his tracts on the subject demonstrate is that early Protestant thinking about matrimony contained the seeds of our own more radical individualism.

These days, few would disagree that respect and affection are central to a successful marriage. But most of us would add another ingredient, which had long been viewed skeptically as a reason to wed: romantic love. Burton, in his "Anatomy of Melancholy"—the most widely read book of the seventeenth century after the Bible—reflected a common view when he described marriage as one of several "remedies of love," which was itself an illness to be overcome. Not until the confessional diaries and novels of the late eighteenth and early nineteenth centuries started to influence bourgeois notions of what Jane Austen called "connubial felicity" did romance begin its steady ascent in the marital realm. Today, needless to say, the most respectable reason you can give for getting married is that you have fallen in love. We have managed to create an ideal of matrimony that combines both lifetime companionship and the less stable but more intoxicating pleasures of romantic ardor.

Such great expectations of marital happiness belong to a larger history of the Western emphasis on the self. The philosopher Charles Taylor, in an examination of how our attitude toward interior life has changed over the past five hundred years, argues that the trend line runs in one direction: from a self-understanding gained from our place in larger entities—such as a chain of being or divine order— toward purpose discovered from within, through what we consider to be authentic self-expression. This is the distance Western culture has travelled from the church confessional to the therapist's couch. In turn, the choice of whom to marry has become less about satisfying the demands of family and community than about satisfying oneself. When you add the contraceptive and reproductive technologies that have separated sex from procreation, what you have is a model of heterosexual marriage that is grounded in and almost entirely sustained on individual preference. This is a historically peculiar state of affairs, one that would be alien to our ancestors and to most traditional cultures today. And it makes the push for gay marriage inevitable.

If agreement between two people were all that was required, of course, gays and lesbians would have been marrying for some time already. According to the 2000 census, the first to collect such data, there are five hundred and ninety-four thousand same-sex couples living in the United States (and that's no doubt an under-count, given peoples' reluctance to report their sexual orientation to the government). But marriage requires the consent of a third party—namely, the state.

This fact, too, can be traced to the same early Protestants who gave us companionate marriage. As the journalist E. J. Graff tells us, in her newly reissued popular history "What Is Marriage For?" (Beacon; $16), at the time Luther nailed his proclamations to the door, Rome had no requirement that a priest be present at the wedding ceremony, vows spoken in private were sufficient to create a binding marriage. That caused the aristocracy no end of trouble, given that disobedient children were liable to threaten carefully negotiated contracts by running into the closet with a servant and whispering sweet promises. For the northern European Protestants, the solution was to require, for the first time, a public ceremony with the presence of witnesses. They also transferred the power to enforce the new rules to the emerging secular states. In England, precaution had long taken the form of "banns": a couple's intention to wed was proclaimed in church on three successive Sundays, thus giving a community plenty of time to determine whether either party was committed elsewhere. The state-issued marriage licenses we employ today originated in magistrates' offices and enabled well-to-do families to avoid the banns by attesting to the fitness of the parties in a document.

Informal arrangements nevertheless persisted among those who were less well off, or who lived in less regulated societies. In the early days of the American Republic, with a population scattered across the continent, there were simply too few ministers and justices of the peace to go around. Largely for the practical reason of not wanting to declare so many children bastards, most state courts recognized common-law marriage, established by mutual consent and cohabitation. If you said you were married and the neighbors tended to believe you, then in the eyes of the law you were. (In eleven states, including Texas, this is still the case.) In the decades after the Civil War, however, government bureaucracy, spurred by the moralists and social scientists of the Progressive movement, began to regulate marriage far more aggressively. For the first time, people who sought to wed had to submit to medical examinations, and those with syphilis or gonorrhea were prevented from marrying by criminal statute. By the century's end, state legislatures had all but mandated that couples obtain a license to marry.

These government-sponsored contracts are at the center of the fight over same-sex unions. In the modern administrative state, civil marriage condenses within a single document a vast array of legal, financial, and medical rights and benefits. Like citizenship for the immigrant, it is a passport to a more secure world. As the old status of marriage has come to include such a range of contractual benefits, the debates over equality within and access to the institution have intensified.

Owing to the reforms of the past forty years, men and women now enter the married state with more legal parity than ever before. Under the old doctrine of coverture, a man owned not only his wife's property but her body as well. Today, in nearly every state, men's and women's rights and obligations in alimony, child custody, child support, and property division in divorce have been made formally gender-neutral. Arrests and prosecutions for domestic abuse, rare thirty years ago, are now routine. As recently as 1984, a man could not be prosecuted for raping his own wife; today, it's a crime in all fifty states.

During the same period that the old patriarchal rules were being revised, the Supreme Court struck down a series of laws limiting the right of individuals to marry in the first place. In the most famous case, Loving v. Virginia, a unanimous Court held that anti-miscegenation laws were unconstitutional. "The freedom to marry has long been recognized as one of the vital personal rights essential to the orderly pursuit of happiness by free men," Chief Justice Warren wrote in 1967. A decade later, a Wisconsin law preventing people from marrying if they were behind on their child-support payments was overturned as too burdensome to the "basic civil rights of man." In 1987, the Rehnquist Court deemed the freedom to marry so fundamental that it could not be denied to prison inmates, whose other constitutional rights are routinely abrogated.

Those who opposed extending this right to same-sex couples used to cite the fact that many states outlawed sodomy: how could you sanction the marriage of people who could be arrested for what they did together in the bedroom? Then, last summer, in Lawrence v. Texas, the Court struck down the thirteen remaining sodomy laws in the country, and established a broad constitutional right to sexual privacy. The force and scope of the opinion surprised even its supporters. In a rare gesture, the Court not only over-turned Bowers v. Hardwick, a 1986 opinion upholding a Georgia sodomy statute, but in essence apologized for it: "Its continuance as precedent demeans the lives of homosexual persons." The majority was careful to point out that the decision said nothing about the "formal recognition" of relationships. But opponents of same-sex marriage realized the decision's importance. As Justice Scalia warned in a caustic dissent, "If moral disapprobation of homosexual conduct is 'no legitimate state interest'... what justification could there possibly be for denying the benefits of marriage to homosexual couples exercising 'the liberty protected by the Constitution'? Surely not the encouragement of procreation, since the sterile and the elderly are allowed to marry." Five months later, the Massachusetts high court cited Lawrence in its decision.

Social conservatives now fear that, in the absence of a constitutional amendment to ban gay marriage, the Supreme Court will combine its precedents on the fundamental right to marry with the more recent decisions in favor of gay rights to give same-sex couples their own Loving v. Virginia—nationalizing gay marriage through constitutional interpretation. As a doctrinal matter, such fears are well grounded. Few legal scholars would disagree with Justice Scalia that Lawrence has made this outcome far more plausible.

Politically, however, a gay Loving is unlikely in the foreseeable future. The Supreme Court is seldom a force of social innovation. The first state to strike down its anti-miscegenation law was California, in 1948. For nearly two decades there after, all through the period of Brown v. Board of Education, the Court avoided the controversy over interracial marriage, exercising its discretion not to hear a case. By the time it decided Loving, fourteen states had followed California's lead, and only sixteen still maintained a ban. The Court endorsed what had become a majority position, if not in public opinion at least in legislative fact; the same can be said of Lawrence. The modern conservative obsession with what is selectively dubbed judicial activism has much to do with Roe v. Wade, and the fact that it struck down thirty state laws criminalizing abortion. No less an advocate of abortion rights than Justice Ruth Bader Ginsburg has suggested that it may have been precipitate not to allow the states to come to a stronger consensus before the Court brought the reform process to an end. In the informal electorate of the states, the vote on full civil marriage for same-sex couples now stands at forty-nine to one. It is difficult to conceive of the Supreme Court recognizing a right to gay marriage under these circumstances.

Meanwhile, interest groups on both sides are lobbying legislators and filing more suits, in an attempt to shape the outcome in one jurisdiction at a time. As a legal matter, these are contests over the expansion of the rights and obligations of marriage. As a political and cultural matter, they are contests over something less easy to codify: the official recognition of love. Here the two most important developments in the history of marriage—secular regulation and the rise of the romantic-companionate ideal—become intertwined. The state is being asked not only to distribute benefits equally but to legitimate gay people's love and affection for their partners. The gay couples now marrying in Massachusetts want not only the same protections that straight people enjoy but the social status that goes along with the state's recognition of a romantic relationship.

This is the difference between civil unions and marriage: one is a legal certificate and the other is a public endorsement. Not surprisingly, many Americans who might support the first remain uncomfortable with the second. In "Gay Marriage" (Times Books; $22), the journalist Jonathan Rauch means to persuade such people that same-sex marriage will be good not only for gay people but for marriage in general. Rauch is a conservative—how many books garner blurbs from both George Will and Barney Frank?— and his argument for the benefits to gay people is based largely on the social discipline he thinks it would impose: once gay men and lesbians are allowed to wed, society can begin expecting them to do so, as it does straight people. "The gay rights era will be over and the gay responsibility era will begin," he writes. This soft coercion is a civilizing force, because "no other institution has the power to turn narcissism into partnership, lust into devotion, strangers into kin." We shouldn't expect results too soon, however "As with the coming of capitalism to the Soviet empire, so with the coming of marriage to gay culture. Freedom and responsibility take time to learn." With analogies as inviting as this, one wonders whether snuggling gay lovers ought to take a bus tour of Putin's Russia before heading to the altar. Though clearly a true believer in matrimony, Rauch doesn't make it sound like much fun.

The real threat to marriage, he says, is the growth of registered cohabitation, from domestic partnerships to civil unions, which increasingly includes unwed heterosexuals. He warns that we are heading to the day when marriage—straight or gay—will become "merely an item on a mix-and-match menu of lifestyle options, a truffle in the candy box." For Rauch, whose view of marriage is more medicinal than confectionary, letting gays marry will actually help marriage by relieving the pressure to create alternatives.

What's undeniable is that the battle over same-sex marriage arrives at a time of declining participation in the institution itself. The number of marriages performed each year in the United States (2.3 million) is as low as it has ever been relative to the adult population. As Andrew Hacker has pointed out, nearly half of Americans reach the age of thirty without having married, and almost twelve per cent of women and sixteen per cent of men enter their forties still never having wed—the highest percentages in the nations history.

As the numbers wane, though, the fantasy seems to grow more intense. The wedding industry generates at least seventy billion dollars a year in revenue, which is double the earnings of the movie business. Bridal magazines are some of the most profitable on the newsstand. The ersatz courtship of the Bachelor and the Bachelorette has provided some of the highest-rated television programming in recent years. At the same time, the pressures on the partnership that follows the wedding day are enormous: in an era of low civic participation and high economic insecurity, spouses are ever more relied upon as the sole providers of continuity and human solace. A state-sponsored, lifelong, intimate relationship—or the prospect of it—now carries a heavy and often unbearable responsibility for personal happiness.

What effect will allowing men to marry men and women to marry women have on our peculiarly modern venture of marriage? Proponents typically say that it will have hardly any—that there is no shortage of marriage licenses, and all that will happen is that more citizens and their children will have the benefits of existing family law. The opposition argues that one of the organizing institutions of our society will be imperilled.

History suggests that neither view is quite accurate. Despite comparisons to the repeal of miscegenation laws, no other expansion of the marriage franchise—to the sterile, to slaves, or to interracial couples—has required an alteration in the basic definition of the term: the union of a man and woman as husband and wife. To discount this as mere semantics misses what the definition points up: that marriage, through all its incarnations, has been a procedure that assigns people a new identity based on their gender. For centuries, it has been the ceremony that makes males into husbands and females into wives. Until very recently, this meant a lifetime commitment to both the security and the constriction of a well-defined social role. The symbolic danger that gay marriage poses to such an arrangement is obvious. It alters the public meaning of the word by further draining it of its power to reinforce traditional expectations of behavior. What does it mean to be a husband in a world where a man could have one of his own? This is up to each individual couple, one is tempted to say. Fair enough; but the words we use to describe our relationships are shared cultural property. There is no private language. In this sense, granting the word "marriage" to gay couples will eventually affect everyone.

The mistake is to consider the change in meaning particularly drastic. After all, undoing customary expectations for how a husband and wife behave toward each other has been one of the goals of the women's movement since its inception. Rather than an abrupt departure, same-sex marriage is the culmination of a larger and ultimately more consequential change in the nature of marital relations between men and women.

Which is one of the reasons that the opposition to it is so fierce. It has come to symbolize what is, historically speaking, radical about contemporary marriage: the decline of the patriarchal legal structure and the rise of the goal of self-fulfillment. Gay marriage is unsettling, to many, not because it departs from modern meanings of matrimony but because it embodies them.

14 *Lust*: "Hobbesian Unity" and "Disasters"

Simon Blackburn

Which brings us to the heart of the matter, and the issues that separate pessimists about sexual desire from optimists. We said that lust was the active and excited desire for the pleasures of sexual activity, leaving it unsettled what these pleasures are. The best clue comes from the seventeenth-century philosopher Thomas Hobbes, famous for the bleak view of the state of nature as the war of all against all, but who nevertheless wrote:

> The appetite which men call LUST... is a sensual pleasure, but not only that; there is in it also a delight of the mind: for it consisteth of two appetites together, to please, and to be pleased; and the delight men take in delighting, is not sensual, but a pleasure or joy of the mind, consisting in the imagination of the power they have so much to please.

Here things are going well. A pleases B. B is pleased at what A is doing, and A is pleased at B's pleasure. This should please B, and a feedback loop is set up, since that in turn pleases A. The ascent does not go on forever: we cannot separate A being pleased at B being pleased at A being pleased at B being pleased... for very long without losing track. But we can get quite a long way. I desire you, and desire your desire for me. I hope that you desire my desire for your desire, and if things are going well, you do. There are no cross-purposes, hidden agendas, mistakes, or deceptions. Lust here is like making music together, a joint symphony of pleasure and response. There is a pure mutuality, or what I shall call a Hobbesian unity.

Pleasures here are not just bodily sensations, although the body will be playing its part. The "delights of the mind" are pleasures at doing something. These pleasures involve the idea of oneself, but they are not properly called narcissistic. The subject is not centrally pleased at himself or herself, but at the excitement of the other. Admittedly, it is not just at that, but also at the fact that the other is excited by the self; but this is to be secondary to the perceived state of the other. The mutual awarenesses increase as the body takes over, as it becomes flooded with desire. The involuntary nature of sexual arousal is here part of the pleasure, the signal that the other is beginning the process of involuntary surrender to desire. As Thomas Nagel puts it:

> These reactions are perceived, and the perception of them is perceived, and that perception is in turn perceived; at each step the domination of the person by his body is reinforced, and the sexual partner becomes more possessible by physical contact, penetration, and envelopment.

Hobbes helps to answer the question we posed early on, of why the ecstatic finale can be an experience of communion or being at one with someone else. It is so in the same way that successful music-making is a communion. When the string quartet comes to a triumphant end, the players have been responding and adjusting to each other delicately for the entire performance. No wonder there is a sense of communion on completion. Some philosophers have thought of sex as if it were something like an excited conversation, but that implies more control than should be expected. In conversations we can branch out in all directions, and we devote conscious thought to what we say. Such a model misses out the domination by the body. So in general, a better comparison is to music-making, where the reciprocal sensitivities can be more or less unconscious, and also for that matter where difficulties such as timing are perhaps more salient.

Hobbes also explains why the communion in sex has a better chance of being real than communion with the divine. Conversations with the divine tend to be more one-sided, and some of us think it is an illusion that there is a conversation going on at all.

An extremely important point about Hobbesian unity is that it can be what philosophers call "variably realized." That is, as with a conversation, there is no one way of doing it. This is why sex manuals are so dreadful, except perhaps for unfortunates who do not have a clue anyway, and who need the equivalent of *69 Ways To Have a Conversation* (there are even books that are the equivalent of *69 Ways to Have a Conversation with Yourself*, or so one deduces from subtitles such as *The Secret World at Your Fingertips* or *A Hand in the Bush*). This is also why the "scientific" discipline of sexology, the kind of research that culminated in the Kinsey reports, misses the point, in the same way that an analysis of a conversation conducted with stopwatch and calipers would miss the point. It is not the movements, but the thought behind them, that matter to lust. The way the symphony unfolds can be anatomically as various as the partners can desire or manage, and as psychologically various as well.

Unlike Aristophanes' unity, a metaphysical fusion of two distinct persons, Hobbesian unity is not intrinsically impossible, any more than communication is. In conversation and music it is not just that I do something and you do something that conveniently fits it. It is rather that *we* do something together, shown by our alertness to the other, and the adjustments we make in the light of what the other does. Bodily contact may not even be necessary. In the Nausicaa episode in James Joyce's *Ulysses*, Leopold Bloom and Gertie McDowell, eyeing each other across the beach, use each other's perceived excitement to work themselves to their climaxes. Unlike President Clinton, whose standards for having sex with someone were so remarkably high, I should have said that Bloom and Gertie had sex together.

However, there is much that can go wrong. As with conversation, there is the boor (and the bore) and the solipsist who loves only the sound of his own voice. There are people paralyzed by shyness, or who fear to speak because they compare themselves, or dread comparison, with others. There are people who are suspicious, and who cannot interpret each other. And the unity may be achieved only because one partner has been "constructed" or molded by the other, obediently taking pleasure in what the other does regardless of his or her suppressed bent, like the wife caused to pretend to enjoy conversations about football and car mechanics until the time comes when she actually does. But whether even that is a suppression of a "real self" underneath, or the comfortable change to new interests, might be a matter of interpretation. Not all education and change is the loss of a Wordsworthian true and innocent self.

We can imagine we share a Hobbesian unity when we do not actually share one. You can think you have caused reciprocated delight when you haven't, as the first page of *Tristram Shandy* reminds us, when at the very moment of his father's crisis, the moment of impregnation,

> *Pray my dear*, quoth my mother, *have you not forgot to wind up the clock?*———Good G———! cried my father, making an exclamation, but taking care to moderate his voice at the same time,———*Did ever woman, since the creation of the world, interrupt a man with such a silly question?*

Tristram trembles to think what check this must have been to the "animal spirits" and what a sad foundation it must have laid for the growth of the poor dispirited fetus that became him. But then we all know lust *can* go wrong, and its trials and strains are the stuff of humor as well as tragedy. There is a nice cartoon of two somewhat disappointed-looking people in bed: "What's the matter, couldn't you think of anyone else either?"

DISASTERS

We can contrast Hobbesian unity with Immanuel Kant's account of the matter. In a notorious passage, Kant tells us that

> Love, as human affection, is the love that wishes well, is amicably disposed, promotes the happiness of others and rejoices in it. But now it is plain that those who merely have sexual inclination love the person from none of the foregoing motives of true human affection, are quite unconcerned for their happiness, and will even plunge them into the greatest unhappiness, simply to satisfy their own inclination and appetite. Sexual love makes of the loved person an object of appetite; as soon as the other person is possessed, and the appetite sated, they are thrown away "as one throws away a lemon that is sucked dry."

The comparison of the used partner to leftover food was there earlier in Shakespeare. Antony says to Cleopatra: "I found you as a morsel cold upon dead Caesars trencher," and Troilus says of Cressida: "The fragments, scraps, the bits, and greasy relics/Of her o'ereaten faith, are given to Diomed." But these are moments of the quite special disenchantment and disgust that assails us on thinking of a third person being involved with our special partner or even ex-partner or hoped-for partner. It is quite another thing to turn those moments of disgust into the universal aftermath of lust. People do not in general see their recent partners in ecstasy as leftover food, nor expect to be seen that way themselves. Even in a post-coital slump one can go on quietly doting.

In Kant's picture, lust objectifies the other person, using him or her as a mere means, a tool of one's own purposes. It is dehumanizing and degrading, and according to Kant it is morally forbidden, since you may never use another person as a mere means to satisfy your own ends. The other person is reduced to a body part, and indeed Kant calls marriage a contract for each to use the other's genitals, so it is lucky that he never tried it. And as Barbara Herman points out in a tight and compelling analysis of Kant's sexual ethics, if sex is thought of like this, it is most obscure why marriage goes any way toward making the use of other human beings permissible, in Kant's own terms.

Kant could fairly be said to paint an obscene picture of lust, one in which all the emphasis is on body parts, and the human being, the person whose parts they are, becomes relatively invisible. The creature that lusts after Beauty is the Beast. Unhappily, many women, and some men, will recognize his account. Indeed, some think it universal, as Kant does, while others think it is inevitable under social and political conditions in which one partner, usually the male, has more power than the other, leading to an inevitable erasure of the personality of the weaker partner, who becomes just the servant who bears genitals to the service of the other.

Perhaps the most notorious account of lust along these lines is Freud's essay "On the Universal Tendency to Debasement in the Sphere of Love," tracing the way in which the idea of the partner as degraded becomes essential to men's sexual enjoyments. Freud works with an opposition between tender, affectionate feelings on the one hand and sensual feelings on the other. The former originate in affection toward the mother, and remain attached to mothers and sisters and respectable women like them. The latter are diverted from these, their desired choices, by the barriers of the incest taboo, and the disgust, shame, and morality that surround and bolster that taboo. So in order for sex to be any good, the male needs women unlike the mother and sister, degraded women, or women who are acceptably degradable. Men may marry women who resemble their mothers and sisters, but they find mistresses among those degraded women to whom they need ascribe no aesthetic misgivings. For Freud, the full sexual satisfaction that these lower women provide comes from the fact that the man can walk away with his soul "intact and gratified" since, having no aesthetic sense, the woman cannot criticize him. Freud was not to know of the kind of conversation that goes on among women in *Sex and the City*, and his sublime conceit never permitted him to imagine it.

In a nutshell, then, sex is either too disgusting to engage in, or when engaged in, not disgusting enough to be gratifying unless one can make use of one's servants and maids. There is also the parallel problem for women, less emphasized by Freud, which results in a taste for hunky morons, such as coal delivery men or well-hung footmen. Like so much of Freud, this might all sound merely funny until we remember, for instance, how much of the lynch mentality in the southern United States was fueled by white male fears of their women's illicit lust for degrading liaisons with black men. All that is needed for Freud's picture is the idea of sexuality as intrinsically degrading, either to oneself or to whomever one happens to be connected. He may be right that this sad idea was widespread among the Viennese upper middle classes of his time, and that for such minds, lust's only escape was to wallow in the supposed degradation of sex with the lower orders, but he is hardly right that it is, or has to be, universal, any more than the snobbery it trades upon.

Freud at least sees joyous degradation in terms of a kind of human relationship, albeit one reaching tenuously across the almost impenetrable class barrier. In this he is one better than Kant. But rather like medieval confessors cataloging forbidden sexual positions, feminist philosophers have carefully dissected the forms and varieties of objectification. In a classic paper, Martha Nussbaum lists seven features that crisscross and overlap in different ways. First, there is instrumentality—using the other as a mere tool of one's purposes. Then, there is denial of autonomy—treating the other as not having a mind of their own, as lacking in self-determination. Third is inertness—treating the other as passive, as lacking in agency and perhaps also in activity (as with Dijkstra's sleeping household nuns). Fourth is fungibility—treating the other as interchangeable with objects of the same type or other types. Fifth is violability—treating the other as lacking in boundary integrity, or as something it is permissible to violate, break up, smash, or break into. Sixth is ownership— treating the other as something that can be disposed of, bought, or sold. Finally, there is denial of subjectivity—treating the other as something whose experiences and feelings (if any) need not be taken into account. Following Nussbaum, Rae Langton adds the general insensitivity to the real nature of the other, as when the woman's voice is no longer heard, or the rapist takes "no" to mean yes.

There are indeed many ways of going wrong here, and we are right to be on the lookout for them. Even without digging into the darker regions of desire, it is undoubtedly true that they structure much of many people's sexual experience. Nussbaum illustrates the dangers with examples drawn from fiction, but if we are to believe them, such figures as Henry Miller and Norman Mailer, boastfully advertising the brutality of their phallic battering rams, illustrate most of these vices. The rapist illustrates the fifth in a more dangerous way, while the "commodification" of women, often supposed to be an integral element in pornography, is captured in the sixth. Too many men conceive of their sexuality like their mountaineering, in terms of domination and conquest, while doubtless many people of both sexes are insensitive to the desires and pleasures of their partners. There is plenty of room for tears at bedtime.

If men are socially and economically dominant, it may most often be they who objectify women. In the brutal capitalist world, it may become easy to think that everything has a money value, and can be bought and sold. But selfishness and insensitivity are nobody's monopoly, and it can work the other way around. The most elegant, if ironic, literary expression owning up to such a lust is actually by a woman, Edna St. Vincent Millay:

> I, being born a woman and distressed
> By all the needs and notions of my kind,
> Am urged by your propinquity to find
> Your person fair, and feel a certain zest
> To bear your body's weight upon my breast:
> So subtly is the fume of life designed,
> To clarify the pulse and cloud the mind,
> And leave me once again undone, possessed.
> Think not for this, however, the poor treason
> Of my stout blood against my staggering brain,
> I shall remember you with love, or season
> My scorn with pity, —let me make it plain:
> I find this frenzy insufficient reason
> For conversation when we meet again.

There is a more general male anxiety that women objectify men. The seventeenth-century poets Sir Thomas Nashe and John Wilmot, Earl of Rochester, each wrote despairingly (or perhaps mock-despairingly) of the inadequacy of men faced with competition from the dildo, imagining, that is, that women only want to use men for one thing, and that one thing more reliably provided without a person on the other end of it.

Although the items on Nussbaum's list look bad and are bad, unfortunately some of them are close neighbors of things that are quite good. We have already met three of them: the way in which ecstasy takes over other cognitive functionings, the intertwining of love and illusion, and the limitations of Aristophanes' myth. Consider the first. At the time of crisis, it is probably true that lovers are not treating their partners decorously or with respect or as fully self-directed moral agents. But that is because strictly speaking they are not treating them any way at all, either as persons or as objects. In the frenzy they are lost to the world, way beyond that. But that is no cause for complaint; indeed the absence of this feature is more often a disappointment, to either the person who does not get there, the partner, or both. Even Nussbaum, who is very sensitive to context, falters here, talking of the loss of boundaries, the surrender of identity, as objectification. But it is not objectification, because it is not treating the other either in an inappropriate way or in a particularly wonderful way. The player is sufficiently lost in the music to become oblivious even to the other players. The body has taken over, saturated with excitement and desire. But this is marvellous, even if moments of rapture mean a pause in the conversation.

Crystallization and the creation of illusions about the self and the other also border on objectification, as Rae Langton notices. We want to be loved for ourselves, not treated as blank canvases on which a lover inscribes his or her own dreams and fantasies. We are not even comfortable when put on a pedestal. Pedestals restrict movement, and there is a long way to fall. But as we have already discussed, imagination may be integral to love. Others cannot discover what she sees in him or he sees in her, because they do not share the crystallization. We do not mind a bit of this, and if it is integral to love we can drink in quite a lot.

Perhaps we prefer Cupid to have dim sight rather than to be totally blind, but it is also just as well that he is not totally clearsighted.

Imaginings and fantasies can lead people into the kind of play acting when lovers infantilize each other (surely much more common than Freud's allegedly universal degradation). And here again a genuine distortion and flaw may be quite close to something that is a harmless part of the repertoire. Intimate behavior is quite often infantile. Lovers are silly. They tease and giggle and tickle each other, and they use childish endearments. We talk of love play, and sex toys and romps, and play it often is. On Valentine's Day, newspapers in Britain are full of personal advertisements along the lines of "Pooh loves Piglet, yum, yum." These may offend against good taste, but they are scarcely a problem for the moralist. A theatrical performance of being less than a full adult, and therefore happily dependent upon the other, seems to be a perfectly legitimate signal of private trust. It displays that you can put yourself in the other person's hands, let your guard down, and throw your dignity to the winds, and yet feel perfectly safe. The same might be said for more lurid actings-out of scenarios of domination and surrender, in which case the bondage gear of the pop concert doesn't answer to anything more sinister than a desire for safety and trust. Perhaps this is confirmed by the femininity of the dominating male.

Such intimacies are properly private. We would be embarrassed at being discovered during them. The intense desire for sexual privacy is frequently misinterpreted as shame at doing something that therefore must be intrinsically shameful or even disgusting. But the desire for privacy should not be moralized like that. Our intimacies are just as private as our couplings. Embarrassment arises because when we are looked upon or overheard by someone else, there is a complete dissonance between what they witness—infantile prattlings, or, if their gaze is obscene, just the twitchings and spasms of the bare forked animals—and the view from the inside, the meanings that are infusing the whole enterprise.

15 Sonnet XXIX, CXVI

William Shakespeare

XXIX

When in disgrace with fortune and men's eyes,
I all alone beweep my outcast state,
And trouble deaf Heaven with my bootless cries,
And look upon myself, and curse my fate,
Wishing me like to one more rich in hope,
Featur'd like him, like him with friends possess'd,
Desiring this man's art, and that man's scope,
With what I most enjoy contented least;
Yet in these thoughts myself almost despising,
Haply I think on thee,—and then my state
(Like to the lark at break of day arising
From sullen earth) sings hymns at heaven's gate;
    For thy sweet love remember'd such wealth brings,
    That then I scorn to change my state with kings'.

CXVI

Let me not to the marriage of true minds
Admit impediments. Love is not love
Which alters when it alteration finds,
Or bends with the remover to remove:
O, no; it is an ever-fixed mark,
That looks on tempests and is never shaken;
It is the star to every wandering bark,
Whose worth's unknown, although his height be taken.
Love's not Time's fool, though rosy lips and cheeks

Within his bending sickle's compass come;
Love alters not with his brief hours and weeks,
But bears it out even to the edge of doom.
If this be error and upon me proved,
I never writ, nor no man ever loved.

16

For Anne Gregory

W.B. Yeats

"Never shall a young man,
Thrown into despair
By those great honey-coloured
Ramparts at your ear,
Love you for yourself alone
And not your yellow hair."

"But I can get a hair-dye
And set such colour there,
Brown, or black, or carrot,
That young men in despair
May love me for myself alone
And not my yellow hair."

"I heard an old religious man
But yesternight declare
That he had found a text to prove
That only God, my dear,
Could love you for yourself alone
And not your yellow hair."

17 *Sex Without Love*

Sharon Olds

How do they do it, the ones who make love
without love? Beautiful as dancers,
gliding over each other like ice-skaters
over the ice, fingers hooked
inside each other's bodies, faces
red as steak, wine, wet as the
children at birth whose mothers are going to
give them away. How do they come to the
come to the come to the God come to the
still waters, and not love
the one who came there with them, light
rising slowly as steam off their joined
skin? These are the true religious,
the purists, the pros, the ones who will not
accept a false Messiah, love the
priest instead of the God. They do not
mistake the lover for their own pleasure,
they are like great runners: they know they are
alone
with the road surface, the cold, the wind,
the fit of their shoes, their over-all cardio
vascular health-just factors, like the partner
in the bed, and not the truth, which is the
single body alone in the universe
against its own best time.

18 *Nicomachean Ethics:* Friendship

Aristotle

9. The Varieties of Friendship
9.1 The Problems
9.11 Common beliefs about friendship

After that the next topic to discuss is friendship; for it is a virtue, or involves virtue, and besides is most necessary for our life.

It is necessary in all external conditions...

For no one would choose to live without friends even if he had all the other goods. For in fact rich people and holders of powerful positions, even more than other people, seem to need friends. For how would one benefit from such prosperity if one had no opportunity for beneficence, which is most often displayed, and most highly praised, in relation to friends? And how would one guard and protect prosperity without friends, when it is all the more precarious the greater it is? In poverty also, and in the other misfortunes, people think friends are the only refuge.

In all times of life...

Moreover, the young need it to keep them from error. The old need it to care for them and support the actions that fail because of weakness. And those in their prime need it, to do fine actions; for 'when two go together . . .', they are more capable of understanding and acting.

And throughout nature...

Further, a parent would seem to have a natural friendship for a child, and a child for a parent, not only among human beings but also among birds and most kinds of animals. Members of the same race, and human beings most of all, have a natural friendship for each other; that is why we praise friends of humanity. And in our travels we can see how every human being is akin and beloved to a human being.

For communities as well as for individuals
Moreover, friendship would seem to hold cities together, and legislators would seem to be more concerned about it than about justice. For concord would seem to be similar to friendship and they aim at concord above all, while they try above all to expel civil conflict, which is enmity.

Further, if people are friends, they have no need of justice; but if they are just they need friendship in addition; and the justice that is most just seems to belong to friendship.

It is both necessary and fine
However, friendship is not only necessary, but also fine. For we praise lovers of friends, and having many friends seems to be a fine thing. Moreover, people think that the same people are good and also friends.

9.12 *Puzzles about friendship*

Still, there are quite a few disputed points about friendship for some hold it is a sort of similarity and that similar people are friends. Hence the saying 'Similar to similar', and 'Birds of a feather', and so on. On the other hand it is said that similar people are all like the proverbial potters, quarrelling with each other.

On these questions some people inquire at a higher level, more proper to natural science. Euripides says that when earth gets dry it longs passionately for rain, and the holy heaven when filled with rain longs passionately to fall into the earth; and Heracleitus says that the opponent cooperates, the finest harmony arises from discordant elements, and all things come to be in struggle. Others, e.g. Empedocles, oppose this view, and say that similar aims for similar.

Let us, then, leave aside the puzzles proper to natural science, since they are not proper to the present examination; and let us examine the puzzles that concern human [nature], and bear on characters and feelings.

For instance, does friendship arise among all sorts of people, or can people not be friends if they are vicious?

Is there one species of friendship, or are there more? Some people think there is only one species because friendship allows more and less. But here their confidence rests on an inadequate sign; for things of different species also allow more and less.

9.2 *General Account of Friendship*

9.21 *The object of friendship: What is lovable*

Perhaps these questions will become clear once we find out what it is that is lovable. For, it seems, not everything is loved, but [only] what is lovable, and this is either good or pleasant or useful. However, it seems that what is useful is the source of some good or some pleasure; hence what is good and what is pleasant are lovable as ends.

Do people love what is good, or what is good for them? For sometimes these conflict; and the same is true of what is pleasant. Each one, it seems, loves what is good for him; and while what is good is lovable unconditionally, what is lovable for each one is what is good for him. In fact each one loves not what is good for him, but what appears good for him; but this will not matter, since [what appears good for him] will be what appears lovable. Hence there are these three causes of love.

9.22 Necessary conditions for friendship

There is no friendship for soulless things

Love for a soulless thing is not called friendship, since there is no mutual loving, and you do not wish good to it. For it would presumably be ridiculous to wish good things to wine; the most you wish is its preservation so that you can have it. To a friend, however, it is said, you must wish goods for his own sake.

Friendship is not mere goodwill. . .

If you wish good things in this way, but the same wish is not returned by the other, you would be said to have [only] goodwill for the other. For friendship is said to be reciprocated goodwill.

And not mere reciprocated goodwill

But perhaps we should add that friends are aware of the reciprocated goodwill. For many a one has goodwill to people whom he has not seen but supposes to be decent or useful, and one of these might have the same goodwill towards him. These people, then, apparently have goodwill to each other, but how could we call them friends when they are unaware of their attitude to each other?

Hence, [to be friends] they must have goodwill to each other, wish goods and be aware of it, from one of the causes mentioned above.

9.3 The Three Types of Friendship

9.31 Complete and incomplete species of friendship correspond to the different objects

Now since these causes differ in species, so do the types of loving and types of friendship. Hence friendship has three species, corresponding to the three objects of love. For each object of love has a corresponding type of mutual loving, combined with awareness of it, and those who love each other wish goods to each other in so far as they love each other.

9.32 Friendships for utility and pleasure are incomplete

Those who love each other for utility love the other not in himself, but in so far as they gain some good for themselves from him. The same is true of those who love for pleasure; for they like a witty person not because of his character, but because he is pleasant to themselves.

And so those who love for utility or pleasure are fond of a friend because of what is good or pleasant for themselves, not in so far as the beloved is who he is, but in so far as he is useful or pleasant.

Hence these friendships as well [as the friends] are coincidental, since the beloved is loved not in so far as he is who he is, but in so far as he provides some good or pleasure. And so these sorts of friendships are easily dissolved, when the friends do not remain similar [to what they were]; for if someone is no longer pleasant or useful, the other stops loving him.

9.33 Friendship for utility

What is useful does not remain the same, but is different at different times. Hence, when the cause of their being friends is removed, the friendship is dissolved too, on the assumption that the friendship aims at these [useful results]. This sort of friendship seems to arise especially among older people, since at that age they pursue what is advantageous, not what is pleasant, and also among those in their prime or youth who pursue what is expedient.

Nor do such people live together very much. For sometimes they do not even find each other pleasant. Hence they have no further need to meet in this way if they are not advantageous [to each other]; for each finds the other pleasant [only] to the extent that he expects some good from him. The friendship of hosts and guests is taken to be of this type too.

9.34 Friendships for pleasure

The cause of friendship between young people seems to be pleasure. For their lives are guided by their feelings, and they pursue above all what is pleasant for themselves and what is near at hand. But as they grow up [what they find] pleasant changes too. Hence they are quick to become friends, and quick to stop; for their friendship shifts with [what they find] pleasant, and the change in such pleasure is quick. Young people are prone to erotic passion, since this mostly follows feelings, and is caused by pleasure; that is why they love and quickly stop, often changing in a single day.

These people wish to spend their days together and to live together; for this is how they gain [the good things] corresponding to their friendship.

9.35 Complete friendship is the friendship of good people

But complete friendship is the friendship of good people similar in virtue; for they wish goods in the same way to each other in so far as they are good, and they are good in themselves. [Hence they wish goods to each other for each other's own sake.] Now those who wish goods to their friend for the friend's own sake are friends most of all; for they have this attitude because of the friend himself, not coincidentally. Hence these people's friendship lasts as long as they are good; and virtue is enduring.

Each of them is both good unconditionally and good for his friend, since good people are both unconditionally good and advantageous for each other. They are pleasant in the same ways too, since good people are pleasant both unconditionally and for each other. [They are pleasant for each other] because each person finds his own actions and actions of that kind pleasant, and the actions of good people are the same or similar.

It is reasonable that this sort of friendship is enduring, since it embraces in itself all the features that friends must have. For the cause of every friendship is good or pleasure, either unconditional or for the lover; and every friendship reflects some similarity. And all the features we have mentioned are found in this friendship because of [the nature of] the friends themselves. For they are similar in this way [i.e. in being good]. Moreover, their friendship also has the other things— what is unconditionally good and what is unconditionally pleasant; and these are lovable most of all. Hence loving and friendship are found most of all and at their best in these friends.

These kinds of friendships are likely to be rare, since such people are few. Moreover, they need time to grow accustomed to each other; for, as the proverb says, they cannot know each other before they have shared the traditional [peck of] salt, and they cannot accept each other or be friends until each appears lovable to the other and gains the other's confidence. Those who are quick to treat each other in friendly ways wish to be friends, but are not friends, unless they are also lovable, and know this. For though the wish for friendship comes quickly, friendship does not.

9.4 Differences and Similarities Between Complete and Incomplete Friendship

9.41 The incomplete friendships resemble the complete

This sort of friendship, then, is complete both in time and in the other ways. In every way each friend gets the same things and similar things from each, and this is what must be true of friends. Friendship for pleasure bears some resemblance to this complete sort, since good people are also pleasant to each other. And friendship for utility also resembles it, since good people are also useful to each other.

Incomplete friendships endure to the extent that they resemble complete friendships

With these [incomplete friends] also, the friendships are most enduring when they get the same thing—e.g. pleasure—from each other, and, moreover, get it from the same source, as witty people do. They must not be like the erotic lover and the boy he loves. For these do not take pleasure in the same things; the lover takes pleasure in seeing his beloved, while the beloved takes pleasure in being courted by his lover. When the beloved's bloom is fading, sometimes the friendship fades too; for the lover no longer finds pleasure in seeing his beloved, while the beloved is no longer courted by the lover.

Many, however, remain friends if they have similar characters and come to be fond of each other's characters from being accustomed to them. Those who exchange utility rather than pleasure in their erotic relations are friends to a lesser extent and less enduring friends. Those who are friends for utility dissolve the friendship as soon as the advantage is removed; for they were never friends of each other, but of what was expedient for them.

9.42 But the character of the friends in complete friendship makes it more enduring

Now it is possible for bad people as well [as good] to be friends to each other for pleasure or utility, for decent people to be friends to base people, and for someone with neither character to be a friend to someone with any character. Clearly, however, only good people can be friends to each other because of the other person himself; for bad people find no enjoyment in one another if they get no benefit.

Moreover, it is only the friendship of good people that is immune to slander. For it is hard to trust anyone speaking against someone whom we ourselves have found reliable for a long time; and among good people there is trust, the belief that he would never do injustice [to a friend], and all the other things expected in a true friendship. But in the other types of friendship [distrust] may easily arise.

9.43 Hence incomplete friendships are friendships only to a limited extent

[These must be counted as types of friendship.] For people include among friends [not only the best type, but] also those who are friends for utility, as cities are—since alliances between cities seem to aim at expediency —and those who are fond of each other, as children are, for pleasure. Hence we must presumably also say that such people are friends, but say that there are more species of friendship than one.

On this view, the friendship of good people in so far as they are good is friendship in the primary way, and to the full extent; and the others are friendships by similarity. They are friends in so far as there is something good, and [hence] something similar to [what one finds in the best kind]; for what is pleasant is good to lovers of pleasure. But these [incomplete] types of friendship are not very regularly combined, and the same people do not become friends for both utility and pleasure. For things that [merely] coincide with each other are not very regularly combined.

Friendship has been assigned, then, to these species. Base people will be friends for pleasure or utility, since they are similar in that way. But good people will be friends because of themselves, since they are friends in so far as they are good. These, then, are friends unconditionally; the others are friends coincidentally and by being similar to these.

9.44 Only complete friendship includes all the right states and activities

Friendship includes both states and activities

Just as with the virtues some people are called good in their state of character, others good in their activity, the same is true of friendship. For some people find enjoyment in each other by living together, and provide each other with good things. Others, however, are asleep or separated by distance, and so are not active in these ways, but are in the state that would result in the friendly activities; for distance does not dissolve the friendship unconditionally, but only its activity. But if the absence is long, it also seems to cause the friendship to be forgotten; hence the saying, 'Lack of conversation has dissolved many a friendship'.

The right activities require pleasure in living together

Older people and sour people do not appear to be prone to friendship. For there is little pleasure to be found in them, and no one can spend his days with what is painful or not pleasant, since nature appears to avoid above all what is painful and to aim at what is pleasant.

Those who welcome each other but do not live together would seem to have goodwill rather than friendship. For nothing is as proper to friends as living together; for while those who are in want desire benefit, blessedly happy people [who want for nothing], no less than the others, desire to spend their days together, since a solitary life fits them least of all. But people cannot spend their time with each other if they are not pleasant and do not enjoy the same things, as they seem to in the friendship of companions.

And only the best kind of friendship includes the appropriate activities and attitudes

It is the friendship of good people that is friendship most of all, as we have often said. For what is lovable and choice-worthy seems to be what is unconditionally good or pleasant, and what is lovable and choiceworthy for each person seems to be what is good or pleasant for him; and both of these make one good person lovable and choiceworthy for another good person.

Loving would seem to be a feeling, but friendship a state. For loving occurs no less towards soulless things, but reciprocal loving requires decision, and decision comes from a state; and what makes [good people] wish good to the beloved for his own sake is their state, not their feeling.

Moreover, in loving their friend they love what is good for themselves; for when a good person becomes a friend he becomes a good for his friend. Each of them loves what is good for himself, and repays in equal measure the wish and the pleasantness of his friend; for friendship is said to be equality. And this is true above all in the friendship of good people.

9.45 The characteristics of friendship are found in friendships for pleasure more than in friendships for utility

Among sour people and older people friendship is found less often, since they are worse tempered and enjoy meeting people less, [and so lack] what seems to be most typical and most productive of friendship. That is why young people become friends quickly, but older people do not, since they do not become friends with people in whom they find no enjoyment—nor do sour people. These people have good will to each other, since they wish goods and give help in time of need; but they scarcely count as friends, since they do not spend their days together or find enjoyment in each other, and these things seem to be above all typical of friendship.

9.46 The number of friends distinguishes the best kind of friendship

No one can have complete friendship for many people, just as no one can have an erotic passion for many at the same time; for [complete friendship, like erotic passion,] is like an excess, and an excess is naturally directed at a single individual. Moreover, just as it is hard for the same person to please many people intensely at the same time, it is also hard, presumably, to be good towards many people at the same time.

Besides, he must gain experience of the other too, and become accustomed to him, which is very difficult.

It is possible, however, to please many people when the friendship is for utility or pleasure, since many people can be pleased in these ways, and the services take little time.

9.47 Friendship for pleasure contrasted with friendship for utility

Of these other two types of friendship the friendship for pleasure is more like [real] friendship; for they get the same thing from each other, and they find enjoyment in each other or in the same things. This is what friendships are like among young people; for a generous [attitude] is found here more [than among older people], whereas it is mercenary people who form friendships for utility.

Moreover, blessedly happy people have no need of anything useful, but do need sources of pleasure. For they want to spend their lives with companions, and though what is painful is borne for a short time, no one could continuously endure even The Good Itself if it were painful to him; hence they seek friends who are pleasant. But, presumably, they must also seek friends who are good as well [as pleasant], and good for them too; for then they will have everything that friends must have.

9.48 Evidence of the difference between the two incomplete types of friendship

Someone in a position of power appears to have separate groups of friends; for some are useful to him, others pleasant, but the same ones are not often both. For he does not seek friends who are both pleasant and virtuous, or useful for fine actions, but seeks one group to be witty, when he pursues pleasure, and the other group to be clever in carrying out instructions; and the same person rarely has both features.

Though admittedly, as we have said, an excellent person is both pleasant and useful, he does not become a friend to a superior [in power and position] unless the superior is also superior in virtue; otherwise he does not reach [proportionate] equality by having a proportionate superior. And this superiority both in power and in virtue is not often found.

9.49 Summary: The extent to which incomplete friendships are friendships

The friendships we have mentioned involve equality, since both friends get the same and wish the same to each other, or exchange one thing for another, e.g. pleasure for benefit. But, as we have said, they are friendships to a lesser extent, and less enduring. Because they are both similar and dissimilar to the same thing they seem both to be and not to be friendships. For in so far as they are similar to the friendship of virtue, they appear to be friendships; for that type of friendship includes both utility and pleasure, and one of these types includes utility, the other pleasure. On the other hand, since the friendship of virtue is enduring and immune to slander, while these change quickly and differ from it in many other ways as well, they do not appear to be friendships, in so far as they are dissimilar to that [best] type.

19 *The Gay Science*

Friedrich Nietzsche

14 - *The good man.*
Better a whole-hearted feud
Than a friendship that is glued.

192 - *The good-natured.*
What is the difference between those good-natured people whose faces radiate good will, and other people? They are cheered by the sight of another person and quickly fall in love with him; therefore they are well disposed toward him, and their first judgment is: "I like him." What distinguishes these people is a rapid succession of the following states: the wish to appropriate (they do not scruple over the worth of the other person), quick appropriation, delight in their new possession, and action for the benefit of their latest conquest.

334 - *One must learn to love.*
This is what happens to us in music: First one has to *learn to hear* a figure and melody at all, to detect and distinguish it, to isolate it and delimit it as a separate life. Then it requires some exertion and good will to *tolerate* it in spite of its strangeness, to be patient with its appearance and expression, and kindhearted about its oddity. Finally there comes a moment when we are *used* to it, when we wait for it, when we sense that we should miss it if it were missing; and now it continues to compel and enchant us relentlessly until we have become its humble and enraptured lovers who desire nothing better from the world than it and only it.

But that is what happens to us not only in music. That is how we have *learned to love* all things that we now love. In the end we are always rewarded for our good will, our patience, fairmindedness, and gentleness with what is strange; gradually, it sheds its veil and turns out to be a new and indescribable beauty. That is its *thanks* for our hospitality. Even those who love themselves will have learned it in this way; for there is no other way. Love, too, has to be learned.

20 *Analects:* Book I

Confucius

1. The Master said, 'Is it not a pleasure, having learned something, to try it out at due intervals? Is it not a joy to have friends come from afar? Is it not gentlemanly not to take offence when others fail to appreciate your abilities?'
2. Yu Tzu[1] said, 'It is rare for a man whose character is such that he is good as a son and obedient as a young man to have the inclination to transgress against his superiors; it is unheard of for one who has no such inclination to be inclined to start a rebellion. The gentleman devotes his efforts to the roots, for once the roots are established, the Way will grow therefrom. Being good as a son and obedient as a young man is, perhaps, the root of a man's character. [i][2]
3. The Master said, 'It is rare, indeed, for a man with cunning words and an ingratiating face to be benevolent.'
4. Tseng Tzu said, 'Every day I examine myself on three counts. In what I have undertaken on another's behalf, have I failed to do my best? In my dealings with my friends have I failed to be trustworthy in what I say? Have I passed on to others anything that I have not tried out myself?'
5. The Master said, 'In guiding a state of a thousand chariots, approach your duties with reverence and be trustworthy in what you say; avoid excesses in expenditure and love your fellow men; employ the labour of the common people only in the right seasons.'
6. The Master said, 'A young man should be a good son at home and an obedient young man abroad, sparing of speech but trustworthy in what he says, and should love the multitude at large but cultivate the friendship of his fellow men.[3] If he has any energy to spare from such action, let him devote it to making himself cultivated.'
7. Tzu-hsia said, 'I would grant that a man has received instruction who appreciates men of excellence where other men appreciate beautiful women, who exerts himself to the utmost in the service of his parents and offers his person to the service of his lord, and who, in his

dealings with his friends, is trustworthy in what he says, even though he may say that he has never been taught.'

8. The Master said, 'A gentleman who lacks gravity does not inspire awe. A gentleman who studies is unlikely to be inflexible.'

'Make[4] it your guiding principle to do your best for others and to be trustworthy in what you say. Do not accept as friend anyone who is not as good as you.'

'When you make a mistake, do not be afraid of mending your ways.'

9. Tseng Tzu said, 'Conduct the funeral of your parents with meticulous care and let not sacrifices to your remote ancestors be forgotten, and the virtue of the common people will incline towards fullness.'

10. Tzu-ch'in asked Tzu-kung, 'When the Master arrives in a state, he invariably gets to know about its government. Does he seek this information? Or is it given him?'

Tzu-kung said, 'The Master gets it through being cordial, good, respectful, frugal and deferential. The way the Master seeks it is, perhaps, different from the way other men seek it.'

11. The Master said, 'Observe what a man has in mind to do when his father is living, and then observe what he does when his father is dead. If, for three years, he makes no changes to his father's ways, he can be said to be a good son.'[5]

12. Yu Tzu said,' Of the things brought about by the rites, harmony is the most valuable. Of the ways of the Former Kings, this is the most beautiful, and is followed alike in matters great and small, yet this will not always work: to aim always at harmony without regulating it by the rites simply because one knows only about harmony will not, in fact, work.'

13. Yu Tzu said, 'To be trustworthy in word is close to being moral in that it enables one's words to be repeated.[6] To be respectful is close to being observant of the rites in that it enables one to stay clear of disgrace and insult. If, in promoting good relationship with relatives by marriage, a man manages not to lose the good will of his own kinsmen, he is worthy of being looked up to as the head of the clan.'[7]

14. The Master said, 'The gentleman seeks neither a full belly nor a comfortable home. He is quick in action but cautious in speech.[8] He goes to men possessed of the Way to be put right. Such a man can be described as eager to learn.'

15. Tzu-kung said, '"Poor without being obsequious, wealthy without being arrogant." What do you think of this saying?'

The Master said, '"That will do, but better still "Poor yet delighting in the Way, wealthy yet observant of the rites."'

Tzu-kung said, 'The *Odes* say,

> Like bone cut, like horn polished,
> Like jade carved, like stone ground.

Is not what you have said a case in point?'

16. The Master said, 'Ssu, only with a man like you can one discuss the *Odes*. Tell such a man something and he can see its relevance to what he has not been told.'

The Master said, 'It is not the failure of others to appreciate your abilities that should trouble you, but rather your failure to appreciate theirs.'

End notes

1. For names of persons and places see Glossary.
2. Numbers in square brackets refer to Textual Notes on p. 234.
3. For the contrast of *jen* (fellow men) with *chung* (multitude) see XV11.6 and for a discussion of the wordy *jen* see Introduction pp. 14 and 17.
4. The whole of what follows is found also in IX.25 while the opening sentence is found also in XII.10.
5. This sentence is found again in IV.20. Cf. also XIX,18.
6. For a discussion of the interpretation of this sentence see D. C. Lau, 'On the expression *fu yen*', *Bulletin of the School of Oriental and African Studies*, XXXVI, 2, (1973)1 PP- 424-33.
7. The sense of this last sentence is rather obscure. The present translation, though tentative, is based on a comment of Cheng Hsuan's on the word yin to the *Chou li* (*Chou li chu shu*, 10.24b).
8. cf. IV.24.

21 The Reasons of Love

Harry Frankfurt

1 There has recently been quite a bit of interest among philosophers in issues concerning whether our conduct must invariably be guided strictly by universal moral principles, which we apply impartially in all situations, or whether favoritism of one sort or another may sometimes be reasonable. In fact, we do not always feel that it is necessary or important for us to be meticulously evenhanded. The situation strikes us differently when our children, or our countries, or our most cherished personal ambitions are at stake. We commonly think that it is appropriate, and perhaps even obligatory, to favor certain people over others who may be just as worthy but with whom our relationships are more distant. Similarly, we often consider ourselves entitled to prefer investing our resources in projects to which we happen to be especially devoted, instead of in others that we may readily acknowledge to have somewhat greater inherent merit. The problem with which philosophers have been concerned is not so much to determine whether preferences of this kind are ever legitimate. Rather, it is to explain under what conditions and in what way they may be justified.

An example that has been widely discussed in this connection has to do with a man who sees two people on the verge of drowning, who can save only one, and who must decide which of the two he will try to save. One of them is a person whom he does not know. The other is his wife. It is hard to think, of course, that the man should make up his mind by just tossing a coin. We are strongly inclined to believe that it would be far more appropriate for him, in a situation like this one, to put aside considerations of impartiality or fairness altogether. Surely the man should rescue his wife. But what is his warrant for treating the two endangered people so unequally? What acceptable principle can the man invoke that would legitimate his decision to let the stranger drown?

One of the most interesting contemporary philosophers, Bernard Williams, suggests that the man already goes wrong if he thinks it is incumbent upon him even to look for a principle from which he could infer that, in circumstances like those in which he finds himself, it is permissible to save one's wife. Instead, Williams says, "it might. .. [be] hoped ... that his motivating thought, fully spelled out, would be [just] the thought that it was his wife." If he adds to this the further thought that in situations of this kind it is *permissible* to save one's wife, Williams admonishes that the man is having "one thought too many." In other words, there is something fishy about the whole notion that when his wife is drowning, the man needs to rely upon some general rule from which a reason that justifies his decision to save her can be derived.[1]

2 I am very sympathetic to Williams's line of thought.[2] However, the example as he presents it is significantly out of focus. It cannot work in the way that he intends, if what it stipulates concerning one of the drowning people is merely that she is the man's wife. After all, suppose that for quite good reasons the man detests and fears his wife. Suppose that she detests him too, and that she has recently engaged in several viciously determined attempts to murder him. Or suppose that it was nothing but a cold-bloodedly arranged marriage of convenience anyhow, and that they have never even been in the same room together except during a perfunctory two-minute wedding ceremony thirty years ago. Surely, to specify nothing more than a bare legal relationship between the man and the drowning woman misses the point.

Let us put aside the matter of their civil status, then, and stipulate instead that the man in the example *loves* one (and not the other) of the two people who are drowning. In that case, it certainly would be incongruous for him to look for a reason to save her. If he does truly love her, then he necessarily already has that reason. It is simply that she is in trouble and needs his help. Just in itself, the fact that he loves her entails that he takes her distress as a more powerful reason for going to her aid than for going to the aid of someone about whom he knows nothing. The need of his beloved for help provides him with this reason, without requiring that he think of any additional considerations and without the interposition of any general rules.

To take such things into account would indeed be to have one thought too many. If the man does not recognize the distress of the woman he loves as a reason for saving her rather than the stranger, then he does not genuinely love her at all. Loving someone or something essentially *means* or *consists in*, among other things, taking its interests as reasons for acting to serve those interests. Love is itself, for the lover, a source of reasons. It creates the reasons by which his acts of loving concern and devotion are inspired.[3]

3 Love is often understood as being, most basically, a response to the perceived worth of the beloved. We are moved to love something, on this account, by an appreciation of what we take to be its exceptional inherent value. The appeal of that value is what captivates us and turns us into lovers. We begin loving the things that we love because we are struck by their value, and we continue to love them for the sake of their value. If we did not find the beloved valuable, we would not love it.

This may well fit certain cases of what would commonly be identified as love. However, the sort of phenomenon that I have in mind when referring here to love is essentially something else. As I am construing it, love is not necessarily a response grounded in awareness of the inherent value of its object. It may sometimes arise like that, but it need not do so.

Love may be brought about—in ways that are poorly understood—by a disparate variety of natural causes. It is entirely possible for a person to be caused to love something without noticing its value, or without being at all impressed by its value, or despite recognizing that there really is nothing especially valuable about it. It is even possible for a person to come to love something despite recognizing that its inherent nature is actually and utterly bad. That sort of love is doubtless a misfortune. Still, such things happen.

It is true that the beloved invariably is, indeed, valuable to the lover. However, perceiving that value is not at all an indispensable *formative* or *grounding* condition of the love. It need not be a perception of value in what he loves that moves the lover to love it. The truly essential relationship between love and the value of the beloved goes in the opposite direction. It is not necessarily as a *result* of recognizing their value and of being captivated by it that we love things. Rather, what we love necessarily *acquires* value for us *because* we love it. The lover does invariably and necessarily perceive the beloved as valuable, but the value he sees it to possess is a value that derives from and that depends upon his love.

Consider the love of parents for their children. I can declare with unequivocal confidence that I do not love my children because I am aware of some value that inheres in them independent of my love for them. The fact is that I loved them even before they were born—before I had any especially relevant information about their personal characteristics or their particular merits and virtues. Furthermore, I do not believe that the valuable qualities they do happen to possess, strictly in their own rights, would really provide me with a very compelling basis for regarding them as having greater worth than many other possible objects of love that in fact I love much less. It is quite clear to me that I do not love them more than other children because I believe they are better.

At times, we speak of people or of other things as "unworthy" of our love. Perhaps this means that the cost of loving them would be greater than the benefit of doing so; or perhaps it means that to love those things would be in some way demeaning. In any case, if I ask myself whether my children are worthy of my love, my emphatic inclination is to reject the question as misguided. This is not because it goes so clearly without saying that my children *are* worthy. It is because my love for them is not at all a response to an evaluation either of them or of the consequences for me of loving them. If my children should turn out to be ferociously wicked, or if it should become apparent that loving them somehow threatened my hope of leading a decent life, I might perhaps recognize that my love for them was regrettable. But I suspect that after coming finally to acknowledge this, I would continue to love them anyhow.

It is not because I have noticed their value, then, that I love my children as I do. Of course, I do perceive them to have value; so far as I am concerned, indeed, their value is beyond measure. That, however, is not the basis of my love. It is really the other way around. The particular value that I attribute to my children is not inherent in them but depends upon my love for them. The reason they are so precious to me is simply that I love them so much. As for why it is that human beings do tend generally to love their children, the explanation presumably lies in the evolutionary pressures of natural selection. In any case, it is plainly *on account of* my love for them that they have acquired in my eyes a value that otherwise they would certainly not possess.

This relationship between love and the value of the beloved—namely, that love is not necessarily grounded in the value of the beloved but does necessarily make the beloved valuable to the lover—holds not only for parental love but quite generally.[4] Most profoundly, perhaps, it is love that accounts for the value to us of life itself. Our lives normally have for us a value that we accept as commandingly authoritative. Moreover, the value to us of living radiates pervasively. It radically conditions the value that we attribute to many other things. It is a powerful—indeed, a comprehensively foundational—generator of value. There are innumerable things that we care about a great deal, and that therefore are very important to us, just because of the ways in which they bear upon our interest in survival.

Why do we so naturally, and with such unquestioning assurance, take self-preservation to be an incomparably compelling and legitimate reason for pursuing certain courses of action? We certainly do not assign this overwhelming importance to staying alive because we believe that there is some great value inherent in our lives, or in what we are doing with them—a value that is independent of our own attitudes and dispositions. Even when we think rather well of ourselves, and suppose that our lives may actually be valuable in such a way, that is not normally why we are so determined to hang on to them. We take the fact that some course of action would contribute to our survival as a reason for pursuing it just because, presumably again thanks to natural selection, we are innately constituted to love living.

4 Let me now attempt to explain what I have in mind when I speak here of love.

The object of love is often a concrete individual: for instance, a person or a country. It may also be something more abstract: for instance, a tradition, or some moral or nonmoral ideal. There will frequently be greater emotional color and urgency in love when the beloved is an individual than when it is something like social justice, or scientific truth, or the way a certain family or a certain cultural group does things; but that is not always the case. In any event, it is not among the defining features of love that it must be hot rather than cool.

One distinctive feature of loving has to do with the particular status of the value that is accorded to its objects. Insofar as we care about something at all, we regard it as important to ourselves; but we may consider it to have that importance only because we regard it as a means to something else. When we love something, however, we go further. We care about it not as merely a means, but as an end. It is in the nature of loving that we consider its objects to be valuable in themselves and to be important to us for their own sakes.

Love is, most centrally, a *disinterested* concern for the existence of what is loved, and for what is good for it. The lover desires that his beloved flourish and not be harmed; and he does not desire this just for the sake of promoting some other goal. Someone might care about social justice only because it reduces the likelihood of rioting; and someone might care about the health of another person just because she cannot be useful to him unless she is in good shape. For the lover, the condition of his beloved is important in itself, apart from any bearing that it may have on other matters.

Love may involve strong feelings of attraction, which the lover supports and rationalizes with flattering descriptions of the beloved. Moreover, lovers often enjoy the company of their beloveds, cherish various types of intimate connection with them, and yearn for reciprocity. These enthusiasms are not essential. Nor is it essential that a person like what he loves. He may even find it distasteful. As in other modes of caring, the heart of the matter is neither affective nor cognitive. It is volitional. Loving something has less to do with what a person believes, or with how he feels, than with a configuration of the will that consists in a practical concern for what is good for the beloved. This volitional configuration shapes the dispositions and conduct of the lover with respect to what he loves, by guiding him in the design and ordering of his relevant purposes and priorities.

It is important to avoid confusing love—as circumscribed by the concept that I am defining—with infatuation, lust, obsession, possessiveness, and dependency in their various forms. In particular, relationships that are primarily romantic or sexual do not provide very authentic or illuminating paradigms of love as I am construing it. Relationships of those kinds typically include a number of vividly distracting elements, which do not belong to the essential nature of love as a mode of disinterested concern, but that are so confusing that they make it nearly impossible for anyone to be clear about just what is going on. Among relationships between humans, the love of parents for their infants or small children is the species of caring that comes closest to offering recognizably pure instances of love.

There is a certain variety of concern for others that may also be entirely disinterested, but that differs from love because it is impersonal. Someone who is devoted to helping the sick or the poor for their own sakes may be quite indifferent to the particularity of those whom he seeks to help. What qualifies people to be beneficiaries of his charitable concern is not that he loves them. His generosity is not a response to their identities as individuals; it is not aroused by their personal characteristics. It is induced merely by the fact that he regards them as members of a relevant class. For someone who is eager to help the sick or the poor, any sick or poor person will do.

When it comes to what we love, on the other hand, that sort of indifference to the specificity of the object is out of the question. The significance to the lover of what he loves is not that his beloved is an instance or an exemplar. Its importance to him is not generic; it is ineluctably particular. For a person who wants simply to help the sick or the poor, it would make perfectly good sense to choose his beneficiaries randomly from among those who are sick or poor enough to qualify. It does not matter who in particular the needy persons are. Since he does not really care about any of them as such, they are entirely acceptable substitutes for each other. The situation of a lover is very different. There can be no equivalent substitute for his beloved. It might really be all the same to someone moved by charity whether he helps this needy person or that one. It cannot possibly be all the same to the lover whether he is devoting himself disinterestedly to what he actually does love or—no matter how similar it might be—to something else instead.

Finally, it is a necessary feature of love that it is not under our direct and immediate voluntary control. What a person cares about, and how much he cares about it, may under certain conditions be up to him. It may at times be possible for him to bring it about that he cares about something, or that he does not care about it, just by making up his mind one way or the other. Whether the requirements of protecting and supporting that thing provide him with acceptable reasons for acting, and how weighty those reasons are, depends in cases like that upon what he himself decides. With regard to certain things, however, a person may discover that he cannot affect whether or how much he cares about them merely by his own decision. The issue is not up to him at all.

For instance, under normal conditions people cannot help caring quite a bit about staying alive, about remaining physically intact, about not being radically isolated, about avoiding chronic frustration, and so on. They really have no choice. Canvassing reasons and making judgments and decisions will not change anything. Even if they should consider it a good idea to stop caring about whether they have any contact with other human beings, or about fulfilling their ambitions, or about their lives and their limbs, they would be unable to stop. They would find that, whatever they thought or decided, they were still disposed to protect themselves from extreme physical and psychic deprivation and harm. In matters like these, we are subject to a necessity that forcefully constrains the will and that we cannot elude merely by choosing or deciding to do so.[5]

The necessity by which a person is bound in cases like these is not a cognitive necessity, generated by the requirements of reason. The way in which it makes alternatives unavailable is not by limiting, as logical necessities do, the possibilities of coherent thought. When we understand that a proposition is self-contradictory, it is impossible for us to believe it, similarly, we cannot help accepting a proposition when we understand that to deny it would be to embrace a contradiction. What people cannot help caring about, on the other hand, is not mandated by logic. It is not primarily a constraint upon belief. It is a volitional necessity, which consists essentially in a limitation of the will.

There are certain things that people cannot do, despite possessing the relevant natural capacities or skills, because they cannot muster the will to do them. Loving is circumscribed by a necessity of that kind: what we love and what we fail to love is not up to us. Now the necessity that is characteristic of love does not constrain the movements of the will through an imperious surge of passion or compulsion by which the will is defeated and subdued. On the contrary, the constraint operates from within our own will itself. It is by our own will, and not by any external or alien force, that we are constrained. Someone who is bound by volitional necessity is unable to form a determined and effective intention—regardless of what motives and reasons he may have for doing so—to perform (or to refrain from performing) the action that is at issue. If he undertakes an attempt to perform it, he discovers that he simply cannot bring himself to carry the attempt all the way through.

Love comes in degrees. We love some things more than we love others. Accordingly, the necessity that love imposes on the will is rarely absolute. We may love something and yet be willing to harm it, in order to protect something else for which our love is greater. A person may well find it possible under certain conditions, then, to perform an act that under others he could not bring himself to perform. For instance, the fact that a person sacrifices his life when he believes that doing so will save his country from catastrophic harm does not reveal thereby that he does not love living; nor does his sacrifice show that he could also have brought himself to accept death willingly when he believed that there was less to be gained. Even of people who commit suicide because they are miserable, it is generally true that they love living. What they would really like, after all, would be to give up not their lives but their misery.

5 There is among philosophers a recurrent hope that there are certain final ends whose unconditional adoption might be shown to be in some way a requirement of reason. But this is a will-o'-the-wisp.[6] There are no necessities of logic or of rationality that dictate what we are to love. What we love is shaped by the universal exigencies of human life, together with those other needs and interests that derive more particularly from the features of individual character and experience. Whether something is to be an object of our love cannot be decisively evaluated either by any a priori method or through examination of just its inherent properties. It can be measured only against requirements that are imposed upon us by other things that we love. In the end, these are determined for us by biological and other natural conditions, concerning which we have nothing much to say.[7]

The origins of normativity do not lie, then, either in the transient incitements of personal feeling and desire, or in the severely anonymous requirements of eternal reason. They lie in the contingent necessities of love. These move us, as feelings and desires do; but the motivations that love engenders are not merely adventitious or (to use Kant's term) heteronomous. Rather, like the universal laws of pure reason, they express something that belongs to our most intimate and most fundamental nature. Unlike the necessities of reason, however, those of love are not impersonal. They are constituted by and embedded in structures of the will through which the specific identity of the individual is most particularly defined.

Of course, love is often unstable. Like any natural condition, it is vulnerable to circumstance. Alternatives are always conceivable, and some of them maybe attractive. It is generally possible for us to imagine ourselves loving things other than those that we do love, and to wonder whether that might not be in some way preferable. The possibility that there may be superior alternatives does not imply, however, that our behavior is irresponsibly arbitrary when we wholeheartedly adopt and pursue the final ends that our loving actually sets for us. Those ends are not fixed by shallow impulse, or by gratuitous stipulation; nor are they determined by what we merely happen at one time or another to find appealing or to decide that we want. The volitional necessity that constrains us in what we love may be as rigorously unyielding to personal inclination or choice as the more austere necessities of reason. What we love is not up to us. We cannot help it that the direction of our practical reasoning is in fact governed by the specific final ends that our love has defined for us. We cannot fairly be charged with reprehensible arbitrariness, nor with a willful or negligent lack of objectivity, since these things are not under our immediate control at all.

To be sure, it may at times be within our power to control them indirectly. We are sometimes capable of bringing about conditions that would cause us to stop loving what we love, or to love other things. But suppose that our love is so wholehearted, and that we are so satisfied to be in its grip, that we could not bring ourselves to alter it even if measures by which it could be altered were available. In that case, the alternative is not a genuine option. Whether it would be better for us to love differently is a question that we are unable to take seriously. For us, as a practical matter, the issue cannot effectively arise.

6 In the end, our readiness to be satisfied with loving what we actually do love does not rest upon the reliability of arguments or of evidence. It rests upon confidence in ourselves. This is not a matter of being satisfied with the range and reliability of our cognitive faculties, or of believing that our information is sufficient. It is confidence of a more fundamental and personal variety. What ensures that we accept our love without equivocation, and what thereby secures the stability of our final ends, is that we have confidence in the controlling tendencies and responses of our own volitional character.

It is by these nonvoluntary tendencies and responses of our will that love is constituted and that loving moves us. It is by these same configurations of the will, moreover, that our individual identities are most fully expressed and defined. The necessities of a person's will guide and limit his agency. They determine what he may be willing to do, what he cannot help doing, and what he cannot bring himself to do. They determine as well what he may be willing to accept as a reason for acting, what he cannot help considering to be a reason for acting, and what he cannot bring himself to count as a reason for acting. In these ways, they set the boundaries of his practical life; and thus they fix his shape as an active being. Any anxiety or uneasiness that he comes to feel on account of recognizing what he is constrained to love goes to the heart, then, of his attitude toward his own character as a person. That sort of disturbance is symptomatic of a lack of confidence in what he himself is.

The psychic integrity in which self-confidence consists can be ruptured by the pressure of unresolved discrepancies and conflicts among the various things that we love. Disorders of that sort undermine the unity of the will and put us at odds with ourselves. The opposition within the scope of what we love means that we are subject to requirements that are both unconditional and incompatible. That makes it impossible for us to plot a steady volitional course. If our love of one thing clashes unavoidably with our love of another, we may well find it impossible to accept ourselves as we are.

However, it may sometimes happen that there is in fact no conflict among the motivations that our various loves impose upon us, and hence that there is no source or locus within us of opposition to any of them. In that case, we have no basis for any uncertainty or reluctance in acceding to the motivations that our loving engenders. Nothing else that we care about as much, or that is of comparable importance to us, provides a ground for hesitation or doubt. Accordingly, we would be able deliberately to arouse ourselves to resist the requirements of love only by resorting to some contrived ad hoc maneuver. That *would* be arbitrary. On the other hand, it cannot be improperly arbitrary for a person to accept the impetus of a love concerning which he is well informed, and that is coherent with the other demands of his will, for he has no pertinent basis for declining to do so.

End notes

1 Bernard Williams, "Persons, Character and Morality," in his *Moral Luck* (Cambridge University Press, 1981), 18.

2 I do have problems with a couple of the details. For one thing, I cannot help wondering why the man should have even the one thought that it's his wife. Are we supposed to imagine that at first he didn't recognize her? Or are we supposed to imagine that at first he didn't remember that they were married, and had to remind himself of that? It seems to me that the strictly correct number of thoughts for this man is zero. Surely the normal thing is that he sees what's happening in the water, and he jumps in to save his wife. Without thinking at all. In the circumstances that the example describes, any thought whatever is one thought too many.

3 That, precisely, is how love makes the world go 'round.

4 There are certain objects of love—certain ideals, for instance—that do appear in many instances to be loved on account of their value. However, it is not necessary that the love of an ideal originate or be grounded in that way. A person might come to love justice or truth or moral rectitude quite blindly, after all, merely as a result of having been brought up to do so. Moreover, it is generally not considerations of value that account for the fact that a person comes to be selflessly devoted to one ideal or value rather than to some other. What leads people to care more about truth than about justice, or more about beauty than about morality, or more about one religion than about another, is generally not some prior appreciation that what they love more has greater inherent value than what they care about less.

5 If someone under ordinary conditions cared nothing at all about dying or being mutilated, or about being deprived of all human contact, we would not regard him merely as atypical. We would consider him to be deranged. There is no strictly logical flaw in those attitudes, but they count nonetheless as irrational—i.e., as violating a defining condition of humanity. There is a sense of rationality that has very little to do with consistency or with other formal considerations. Thus suppose that a person deliberately causes death or deep suffering for no reason, or (Hume's example) seeks the destruction of a multitude in order to avoid a minor injury to one of his fingers. Anyone who could bring himself to do such things would naturally be regarded—despite his having made no logical error—as "crazy." He would be regarded, in other words, as lacking reason. We are accustomed to understanding rationality as precluding contradiction and incoherence—as limiting what it is possible for us to think. There is also a sense of rationality in which it limits what we can bring ourselves to do or to accept. In the one sense, the alternative to reason is what we recognize as inconceivable. In the other, it is what we find unthinkable.

6 Some philosophers believe that the ultimate warrant for moral principles is to be found in reason. In their view, moral precepts are inescapably authoritative precisely because they articulate conditions of rationality itself. This cannot be correct. The sort of opprobrium that attaches to moral transgressions is quite unlike the sort of opprobrium that attaches to violations of the requirements of reason. Our response to people who behave immorally is not at all the same as our response to people whose thinking is illogical. Manifestly, there is something other than the importance of being rational that supports the injunction to be moral. For a discussion of this point, cf. my "Rationalism in Ethics," in *Autonomes Handeln: Beitrage zur Philosophic von Harry G. Frankfurt*, ed. M. Betzler and B. Guckes (Akademie Verlag, 2000).

7 It may be perfectly reasonable to insist that people *should* care about certain things, which they do not actually care about, but only if something is known about what they *do* in fact care about. If we may assume that people care about leading secure and satisfying lives, for example, we will be justified in trying to see to it that they care about things that we believe are indispensable for achieving security and satisfaction. It is in this way that a "rational" basis for morality may be developed.

22 *The New Testament*: The First Letter of Paul to the Corinthians

Now concerning spiritual gifts, brothers and sisters, I do not want you to be uniformed. You know that when you were pagans, you were enticed and led astray to idols that could not speak. Therefore I want you to understand that no one speaking by the spirit of God ever says "Let Jesus be cursed!" and no one can say "Jesus is Lord" except by the Holy Spirit.

Now there are varieties of gifts, but the same spirit; and there are varieties of service, but the same Lord; and there are varieties of activities, but it is the same God who activates all of them in everyone. To each is given the manifestation of the Spirit for the common good. To one is given through the Spirit the utterance of wisdom, and to another the utterance of knowledge according to the same Spirit, to another faith by the same Spirit, to another to another gifts of healing by the one Spirit, to another the working of miracles, to another prophecy, to another the discernment of spirits, to another various kinds of tongues, to another the interpretation of tongues. All these are activated by one and the same Spirit, who allots to each one individually just as the Spirit chooses.

For just as the body is one and has many members, and all the members of the body, though many, are one body, so it is with Christ. For in the one Spirit we were all baptized into one body-Jews or Greeks, Slaves or Free-and we were all made to drink of one Spirit.

Indeed, the body does not consists of one member but of many. If the foot would say, "Because I am not a hand, I do not belong to the body," that would not make it any less a part of the body. And if the ear would say, "Because I am not an eye, I do not belong to the body," that would not make it any less a part of the body. If the whole body were an eye, where would the hearing be? If the whole body were hearing, where would the sense of smell be? But as it is, God arranged the members of the body, each one of them, as he chose. If all were single member, where would the body be? As it is, there are many members, yet one body. The eye cannot say to the hand, "I have no need of you," nor again the head to the feet, " I have no need of you." On the contrary, the members of the body that seem to be weaker are indispensable, and those members of the body that we think less honorable we clothe with greater honor, and our less respectable members are treated with greater respect; whereas our more respectable members do not need this. But God has so arranged the body, giving the greater honor to the inferior member, that there may be no dissension within the body, but the members may have the same care for one another. If one member suffers, all suffer together with it; if one member is honored, all rejoice together with it.

Now you are the body of Christ and individually members of it. And God has appointed in the church first apostles, second prophets, third teachers; then deeds of power, then gifts of healing, forms of assistance, forms of leadership, various kinds of tongues. Are all apostles? Are all prophets? Are all teachers? Do all work miracles? Do all possess gifts of healing? Do all speak in tongues? Do all interpret? But strive for the greater gifts. And I will show you a still more excellent way.

If I speak in tongues of mortals and of angels, but do not have love, I am a noisy gong or a clanging cymbal. And if I have prophetic powers, and understand all mysteries and all knowledge, and if I have all faith, so as to remove mountains, but do not have love, I am nothing. If I give away all my possessions, and if I hand over my body so that I may boast, but do not have love, I gain nothing.

Love is patient; love is kind; love is not envious or boastful or arrogant or rude. It does not insist on its own way; it it not irritable or resentful; it does not rejoice in wrongdoing, but rejoices in the truth. It bears all things, believes all things, hopes all thing, endures all things.

Love never ends. But as for prophecies, they will come to an end; as for tongues, they will cease; as for knowledge, it will come to an end. For we know only in part, and we prophesy only in part; but when the complete comes, the partial will come to an end. When I was a child, I spoke like a child, I thought like a child, I reasoned like a child; when I became an adult, I put an end to childish ways. For now we see in a mirror, dimly, but then we will see face to face. Now I know only in part; then I will know fully, even as I have been fully known. And now faith, hope, and love abide, these three; and the greatest of these is love.

23 *The Teachings of the Compassionate Buddha:* The Bodhisattva's Vow of Universal Redemption

E.A. Burtt, editor

In these selections, from various sutras, the Mahayana answer to the questions raised in the last two selections of Part III is presented. The Bodhisattva has transcended the state in which he is concerned for his own salvation; he is committed to the eternal weal of all living beings, and will not rest until he has led them all to the goal. On attaining enlightenment he does not leave the world behind and enter Nirvana by himself; he remains in the world, appearing like an ordinary person, but devoting his compassionate skill to the aid of others. He shares and bears the burden of their sufferings, in loving union with them, instead of merely giving others an example of a person who has overcome the causes of suffering for himself.

The Lord: "What do you think, Sariputra, does it occur to any of the Disciples and Pratyekabuddhas to think that 'after we have known full enlightenment, we should lead all beings to Nirvana, into the realm of Nirvana which leaves nothing behind'?"

Sariputra: "No indeed, O Lord."

The Lord: "One should therefore know that this wisdom of the Disciples and Pratyekabuddhas bears no comparison with the wisdom of a Bodhisattva. What do you think, Sariputra, does it occur to any of the Disciples and Pratyekabuddhas that 'after I have practised the six perfections, have brought beings to maturity, have purified the Buddha-field, have fully gained the ten powers of a Tathagata, his four grounds of self-confidence, the four analytical knowledges and the eighteen special dharmas of a Buddha, after I have known full enlightenment, I shall lead countless beings to Nirvana'?"

Sariputra: "No, O Lord."

The Lord: "But such are the intentions of a Bodhisattva. A glowworm, or some other luminous animal, does not think that its light could illuminate the Continent of Jambudvipa, or radiate over it. Just so, the Disciples and Pratyekabuddhas do not think that they should, after winning full enlightenment, lead all beings to Nirvana. But the sun, when it has risen, radiates its light over the whole of Jambudvipa. Just so a Bodhisattva, after he has accomplished the practices which lead to the full enlightenment of Buddhahood, leads countless beings to Nirvana."

The Lord: "Suppose, Subhuti, that there was a most excellent hero, very vigorous, of high social position, handsome, attractive and most fair to behold, of many virtues, in possession of all the finest virtues, of those virtues which spring from the very height of sovereignty, morality, learning, renunciation, and so on. He is judicious, able to express himself, to formulate his views clearly, to substantiate his claims; one who always knows the suitable time, place, and situation for everything. In archery he has gone as far as one can go, he is successful in warding off all manner of attack, most skilled in all arts, and foremost, through his fine achievements, in all crafts. . . . He is versed in all the treatises, has many friends, is wealthy, strong of body, with large limbs, with all his faculties complete, generous to all, dear and pleasant to many. Any work he might undertake he manages to complete, he speaks methodically, shares his great riches with the many, honours what should be honoured, reveres what should be revered, worships what should be worshipped. Would such a person, Subhuti, feel ever-increasing joy and zest?"

Subhuti: "He would, O Lord."

The Lord: "Now suppose, further, that this person, so greatly accomplished, should have taken his family with him on a journey, his mother and father, his sons and daughters. By some circumstance they find themselves in a great, wild forest. The foolish ones among them would feel fright, terror, and hair-raising fear. He, however, would fearlessly say to his family: 'Do not be afraid! I shall soon take you safely and securely out of this terrible and frightening forest. I shall soon set you free!' If then more and more hostile and inimical forces should rise up against him in that forest, would this heroic man decide to abandon his family, and to take himself alone out of that terrible and frightening forest—he who is not one to draw back, who is endowed with all the force of firmness and vigour, who is wise, exceedingly tender and compassionate, courageous and a master of many resources?"

Subhuti: "No, O Lord. For that person, who does not abandon his family, has at his disposal powerful resources, both within and without. On his side forces will arise in that wild forest which are quite a match for the hostile and inimical forces, and they will stand up for him and protect him. Those enemies and adversaries of his, who look for a weak spot, who seek for a weak spot, will not gain any hold over him. He is competent to deal with the situation, and is able, unhurt and uninjured, soon to take out of that forest, both his family and himself, and securely and safely will they reach a village, city, or market town."

The Lord: "Just so, Subhuti, is it with a Bodhisattva who is full of pity and concerned with the welfare of all beings, who dwells in friendliness, compassion, sympathetic joy, and even-mindedness.

"Although the son of Jina[1] has penetrated to this Immutable true nature of dharmas,
Yet he appears like one of those who are blinded by ignorance, subject as he is to birth,
and so on. That is truly wonderful.

It is through his compassionate skill in means for others that he is tied to the world,
And that, though he has attained the state of a saint, yet he appears to be in the state of an ordinary person.

He has gone beyond all that is worldly, yet he has not moved out of the world;
In the world he pursues his course for the world's weal, unstained by worldly taints.

As a lotus flower, though it grows in water, is not polluted by the water,
So he, though born in the world, is not polluted by worldly dharmas.

Like a fire, his mind constantly blazes up into good works for others;
At the same time he always remains merged in the calm of the trances and formless attainments.[2]

Through the power of his previous penetration (into reality), and because he has left all discrimination behind,
He again exerts no effort when he brings living things to maturity.

He knows exactly who is to be educated, how, and by what means,
Whether by his teaching, his physical appearance, his practices, or his bearing.

Without turning towards anything, always unobstructed in his wisdom,
He goes along, in the world of living beings, boundless as space, acting for the weal of beings.

When a Bodhisattva has reached this position, he is like the Tathagatas,
Insofar as he is in the world for the sake of saving beings.

But as a grain of sand compares with the earth, or a puddle in a cow's footprint with the ocean,
So great still is the distance of the Bodhisattvas from the Buddha."

A Bodhisattva resolves: I take upon myself the burden of all suffering, I am resolved to do so, I will endure it. I do not turn or run away, do not tremble, am not terrified, nor afraid, do not turn back or despond.

And why? At all costs I must bear the burdens of all beings. In that I do not follow my own inclinations. I have made the vow to save all beings. All beings I must set free. The whole world of living beings I must rescue, from the terrors of birth, of old age, of sickness, of death and rebirth, of all kinds of moral offence, of all states of woe, of the whole cycle of birth-and-death, of the jungle of false views, of the loss of wholesome dharmas, of the concomitants of ignorance,—from all these terrors I must rescue all beings. ...I walk so that the kingdom of unsurpassed cognition is built up for all beings. My endeavours do not merely aim at my own deliverance. For with the help of the boat of the thought of all-knowledge, I must rescue all these beings from the stream of Samsara, which is so difficult to cross; I must pull them back from the great precipice, I must free them from all calamities, I must ferry them across the stream of Samsara. I myself must grapple with the whole mass of suffering of all beings. To the limit of my endurance I will experience in all the states of woe, found in any world system, all the abodes of suffering. And I must not cheat all beings out of my store of merit. I am resolved to abide in each single state of woe for numberless aeons; and so I will help all beings to freedom, in all the states of woe that may be found in any world system whatsoever.

And why? Because it is surely better that I alone should be in pain than that all these beings should fall into the states of woe. There I must give myself away as a pawn through which the whole world is redeemed from the terrors of the hells, of animal birth, of the world of Yama; and with this my own body I must experience, for the sake of all beings, the whole mass of all painful feelings. And on behalf of all beings I give surety for all beings, and in doing so I speak truthfully, am trustworthy, and do not go back on my word. I must not abandon all beings.

And why? There has arisen in me the will to win all-knowledge, with all beings for its object, that is to say, for the purpose of setting free the entire world of beings. And I have not set out for the supreme enlightenment from a desire for delights, not because I hope to experience the delights of the five sense-qualities, or because I wish to indulge in the pleasures of the senses. And I do not pursue the course of a Bodhisattva in order to achieve the array of delights that can be found in the various worlds of sense-desire.

And why? Truly no delights are all these delights of the world. All this indulging in the pleasures of the senses belongs to the sphere of Mara.

End notes

1 "Conqueror."

2 Insights and indescribable realizations.

24 *Confessions*: Book III, 3-6; Book X, 6-7, 26-29

St. Augustine

BOOK III—3

Yet all the while, far above, your mercy hovered faithfully about me. I exhausted myself in depravity, in the pursuit of an unholy curiosity. I deserted you and sank to the bottom-most depths of scepticism and the mockery of devil-worship. My sins were a sacrifice to the devil, and for all of them you chastised me. I defied you even so far as to relish the thought of lust, and gratify it too, within the walls of your church during the celebration of your mysteries. For such a deed I deserved to pluck the fruit of death, and you punished me for it with a heavy lash. But, compared with my guilt, the penalty was nothing. How infinite is your mercy, my God! You are my Refuge from the terrible dangers amongst which I wandered, head on high, intent upon withdrawing still further from you. I loved my own way, not yours, but it was a truant's freedom that I loved.

Besides these pursuits I was also studying for the law. Such ambition was held to be honourable and I determined to succeed in it. The more unscrupulous I was, the greater my reputation was likely to be, for men are so blind that they even take pride in their blindness. By now I was at the top of the school of rhetoric. I was pleased with my superior status and swollen with conceit. All the same, as you well know, Lord, I behaved far more quietly than the 'Wreckers', a title of ferocious devilry which the fashionable set chose for themselves. I had nothing whatever to do with their outbursts of violence, but I lived amongst them, feeling a perverse sense of shame because I was not like them. I kept company with them and there were times when I found their friendship a pleasure, but I always had a horror of what they did when they lived up to their name. Without provocation they would set upon some timid newcomer, gratuitously affronting his sense of decency for their own amusement and using it as fodder for their spiteful jests. This was the devil's own behavior or not far different. 'Wreckers' was a fit name for them, for they were already adrift and total wrecks themselves. The mockery and trickery which they loved to practise on others was a secret snare of the devil, by which they were mocked and tricked themselves.

4

These were the companions with whom I studied the art of eloquence at that impressionable age. It was my ambition to be a good speaker, for the unhallowed and inane purpose of gratifying human vanity. The prescribed course of study brought me to a work by an author named Cicero, whose writing nearly everyone admires, if not the spirit of it. The title of the book is *Hortensius* and it recommends the reader to study philosophy. It altered my outlook on life. It changed my prayers to you, O Lord, and provided me with new hopes and aspirations. All my empty dreams suddenly lost their charm and my heart began to throb with a bewildering passion for the wisdom of eternal truth. I began to climb out of the depths to which I had sunk, in order to return to you. For I did not use the book as a whetstone to sharpen my tongue. It was not the style of it but the contents which won me over, and yet the allowance which my mother paid me was supposed to be spent on putting an edge on my tongue. I was now in my nineteenth year and she supported me, because my father had died two years before.

My God, how I burned with longing to have wings to carry me back to you, away from all earthly things, although I had no idea what you would do with me! For *yours is the wisdom*[1]. In Greek the word 'philosophy' means 'love of wisdom', and it was with this love that the *Hortensius* inflamed me. There are people for whom philosophy is a means of misleading others, for they misuse its great name, its attractions, and its integrity to give colour and gloss to their own errors. Most of these so-called philosophers who lived in Cicero's time and before are noted in the book. He shows them up in their true colours and makes quite clear how wholesome is the admonition which the Holy Spirit gives in the words of your good and true servant, Paul: *Take care not to let anyone cheat you with his philosophizing, with empty fantasies drawn from human tradition, from worldly principles; they were never Christ's teaching. In Christ the whole plenitude of Deity is embodied and dwells in him.*[2]

But, O Light of my heart, you know that at that time, although Paul's words were not known to me, the only thing that pleased me in Cicero's book was his advice not simply to admire one or another of the schools of philosophy, but to love wisdom itself, whatever it might be, and to search for it, pursue it, hold it, and embrace it firmly. These were the words which excited me and set me burning with fire, and the only check to this blaze of enthusiasm was that they made no mention of the name of Christ. For by your mercy, Lord, from the time when my mother fed me at the breast my infant heart had been suckled dutifully on his name, the name of your Son, my Saviour. Deep inside my heart his name remained, and nothing could entirely captivate me, however learned, however neatly expressed, however true it might be, unless his name were in it.

5

So I made up my mind to examine the holy Scriptures and see what kind of books they were. I discovered something that was at once beyond the understanding of the proud and hidden from the eyes of children. Its gait was humble, but the heights it reached were sublime. It was enfolded in mysteries, and I was not the kind of man to enter into it or bow my head to follow where it led. But these were not the feelings I had when I first read the Scriptures. To me they seemed quite unworthy of comparison with the stately prose of Cicero, because I had too much conceit to accept their simplicity and not enough insight to penetrate their depths. It is surely true that as the child grows these books grow with him. But I was too proud to call myself a child. I was inflated with self-esteem, which made me think myself a great man.

BOOK X—6

My love of you, O Lord, is not some vague feeling: it is positive and certain. Your word struck into my heart and from that moment I loved you. Besides this, all about me, heaven and earth and all that they contain proclaim that I should love you, and their message never ceases to sound in the ears of all mankind, so that there is no excuse for any not to love you. But, more than all this, *you will show pity on those whom you pity; you will show mercy where you are merciful*[3]; for if it were not for your mercy, heaven and earth would cry your praises to deaf ears.

But what do I love when I love my God? Not material beauty or beauty of a temporal order; not the brilliance of earthly light, so welcome to our eyes; not the sweet melody of harmony and song; not the fragrance of flowers, perfumes, and spices; not manna or honey; not limbs such as the body delights to embrace. It is not these that I love when I love my God. And yet, when I love him, it is true that I love a light of a certain kind, a voice, a perfume, a food, an embrace; but they are of the kind that I love in my inner self, when my soul is bathed in light that is not bound by space; when it listens to sound that never dies away; when it breathes fragrance that is not borne away on the wind; when it tastes food that is never consumed by the eating; when it clings to an embrace from which it is not severed by fulfilment of desire. This is what I love when I love my God.

But what is my God? I put my question to the earth. It answered, 'I am not God', and all things on earth declared the same. I asked the sea and the chasms of the deep and the living things that creep in them, but they answered, 'We are not your God. Seek what is above us.' I spoke to the winds that blow, and the whole air and all that lives in it replied, 'Anaximenes[4] is wrong. I am not God.' I asked the sky, the sun, the moon, and the stars, but they told me, 'Neither are we the God whom you seek.' I spoke to all the things that are about me, all that can be admitted by the door of the senses, and I said, ' Since you are not my God, tell me about him. Tell me something of my God.' Clear and loud they answered, 'God is he who made us.' I asked these questions simply by gazing at these things, and their beauty was all the answer they gave.

Then I turned to myself and asked, 'Who are you?' 'A man,' I replied. But it is clear that I have both body and soul, the one the outer, the other the inner part of me. Which of these two ought I to have asked to help me find my God? With my bodily powers I had already tried to find him in earth and sky, as far as the sight of my eyes could reach, like an envoy sent upon a search. But my inner self is the better of the two, for it was to the inner part of me that my bodily senses brought their messages. They delivered to their arbiter and judge the replies which they carried back from the sky and the earth and all that they contain, those replies which stated 'We are not God' and 'God is he who made us'. The inner part of man knows these things through the agency of the outer part. I, the inner man, know these things; I, the soul, know them through the senses of my body. I asked the whole mass of the universe about my God, and it replied, 'I am not God. God is he who made me.'

Surely everyone whose senses are not impaired is aware of the universe around him? Why, then, does it not give the same message to us all? The animals, both great and small, are aware of it, but they cannot inquire into its meaning because they are not guided by reason, which can sift the evidence relayed to them by their senses. Man, on the other hand, can question nature. He is able to *catch sight of God's invisible nature through his creatures*[5], but his love of these material tilings is too great. He becomes their slave, and slaves cannot be judges. Nor will the world supply an answer to those who question it, unless they also have the faculty to judge it. It does not answer in different language - that is, it does not change its aspect - according to whether a man merely looks at it or subjects it to inquiry while he looks. If it did, its appearance would be different in each case. Its aspect is the same in both cases, but to the man who merely looks it says nothing, while to the other it gives an answer. It would be nearer the truth to say that it gives an answer to all, but it is only understood by those who compare the message it gives them through their senses with the truth that is in themselves. For truth says to me, 'Your God is not heaven or earth or any kind of bodily thing.' We can tell this from the very nature of such things, for those who have eyes to see know that their bulk is less in the part than in the whole. And I know that my soul is the better part of me, because it animates the whole of my body. It gives it life, and this is something that no body can give to another body. But God is even more. He is the Life of the life of my soul.

26

Where, then, did I find you so that I could learn of you? For you were not in my memory before I learned of you. Where else, then, did I find you, to learn of you, unless it was in yourself, above me? Whether we approach you or depart from you, you are not confined in any place. You are Truth, and you are everywhere present where all seek counsel of you. You reply to all at once, though the counsel each seeks is different. The answer you give is clear, but not all hear it clearly. All ask you whatever they wish to ask, but the answer they receive is not always what they want to hear. The man who serves you best is the one who is less intent on hearing from you what he wills to hear than on shaping his will according to what he hears from you.

27

I have learnt to love you late, Beauty at once so ancient and so new! I have learnt to love you late! You were within me, and I was in the world outside myself. I searched for you outside myself and, disfigured as I was, I fell upon the lovely things of your creation. You were with me, but I was not with you. The beautiful things of this world kept me far from you and yet, if they had not been in you, they would have had no being at all. You called me; you cried aloud to me; you broke my barrier of deafness. You shone upon me; your radiance enveloped me; you put my blindness to flight. You shed your fragrance about me; I drew breath and now I gasp for your sweet odour. I tasted you, and now I hunger and thirst for you. You touched me, and I am inflamed with love of your peace.

28

When at last I cling to you with all my being, for me there will be no more sorrow, no more toil. Then at last I shall be alive with true life, for my life will be wholly filled by you. You raise up and sustain all whose lives you fill, but my life is not yet filled by you and I am a burden to myself. The pleasures I find in the world, which should be cause for tears, are at strife with its sorrows, in which I should rejoice, and I cannot tell to which the victory will fall. Have pity on me, O Lord, in my misery! My sorrows are evil and they are at strife with joys that are good, and I cannot tell which will gain the victory. Have pity on me, O Lord, in my misery! I do not hide my wounds from you. I am sick, and you are the physician. You are merciful: I have need of your mercy. Is not our life on earth a period of trial? For who would wish for hardship and difficulty? You command us to endure these troubles, not to love them. No one loves what he endures, even though he may be glad to endure it. For though he may rejoice in his power of endurance, he would prefer that there should be nothing for him to endure. When I am in trouble I long for good fortune, but when I have good fortune I fear to lose it. Is there any middle state between prosperity and adversity, some state in which human life is not a trial? In prosperity as the world knows it there is twofold cause for grief, for there is grief in the fear of adversity and grief in joy that does not last. And in what the world knows as adversity the causes of grief are threefold, for not only is it hard to bear, but it also causes us to long for prosperous times and to fear that our powers of endurance may break. Is not man's life on earth a long, unbroken period of trial?

29

There can be no hope for me except in your great mercy. Give me the grace to do as you command, and command me to do what you will! You command us to control our bodily desires. And, as we are told, when I knew that no man can *be master of himself, except of God's bounty, I was wise enough already to know whence the gift came*[6]. Truly it is by continence that we are made as one and regain that unity of self which we lost by falling apart in the search for a variety of pleasures. For a man loves you so much the less if, besides you, he also loves something else which he does not love for your sake. O Love ever burning, never quenched! O Charity, my God, set me on fire with your love! You command me to be continent. Give me the grace to do as you command, and command me to do what you will!

End notes

1 Job 12: 13.

2 Col. 2: 8, 9.

3 I Cor. 13:12.

4 Anaxitnenes of Miletus, the philosopher, who lived in the sixth century B.C. His teaching was that air is the first cause of all things.

5 I Cor. 10:13.

6 Is. 58: 10.

25 *The Life of Saint Teresa by Herself*

St. Teresa of Avila

I have wandered far from my subject. I was trying to explain the reasons why this kind of vision can not be the work of the imagination. For, how could we picture Christ's Humanity merely through having dwelt on it, or compose His great beauty out of our own heads? If such a conception were to be anything like the original, it would take quite a long time to build up. One can indeed construct such a picture from the imagination, and spend quite a while regarding it, and reflecting on the form and brightness of it. One can gradually perfect this picture and commit it to the memory. What is there to prevent this, since it is the work of the intelligence? But when it comes to the visions I am speaking of, there is no way of building them up. We have to look at them when the Lord is pleased to show them to us—to look as He wishes and at what He wishes. We can neither add nor subtract anything, nor can we obtain a vision by any actions of our own. We cannot look at it when we like or refrain from looking at it; if we try to look at any particular feature of it, we immediately lose Christ.

For two and a half years, God granted me this favour at frequent intervals. But more than three years ago He took it from me, in this form of a continual experience, and gave me something of a higher kind, of which I shall perhaps speak later. During all that time, though I saw that He was speaking to me, though I gazed on His very great beauty, and felt the sweetness with which those words of His, which were sometimes stern, issued from His fair and divine mouth, and though, at the same time, I greatly longed to see the colour of His eyes, or His stature, so as to be able to describe them later, I was never worthy enough to see them, nor was it any good my trying to do so. On the contrary, these efforts lost me the vision altogether. Though I sometimes see Him looking at me with compassion, His gaze is so powerful that my soul cannot endure it. It is caught in so sublime a rapture that it loses this lovely vision in order to in crease its enjoyment of the whole. So here there is no question of willingness or unwillingness. It is clear that all the Lord wants of us is humility and shame, and that we shall accept what is given us, with praise for the Giver.

This is true of all visions without exception. There is nothing that we can do about them; no effort of ours makes us see more or less, or calls up or dispels a vision. The Lord desires us to see very clearly that this work is not ours but His Majesty's. We are the less able, therefore, to take pride in it; on the contrary it makes us humble and afraid, when we see that just as the Lord takes away our power of seeing what we will, so He can also remove these favours and His grace, with the result that we are utterly lost. Let us always walk in fear therefore, so long as we are living in this exile.

Almost always Our Lord appeared to me as He rose from the dead, and it was the same when I saw Him in the Host. Only occasionally, to hearten me if I was in tribulation, He would show me His wounds, and then He would appear sometimes on the Cross and sometimes as He was in the Garden. Sometimes too, but rarely, I saw Him wearing the crown of thorns, and sometimes carrying His Cross as well, because of my deeds, let me say, and those of others. But always His body was glorified. Many were the reproaches and trials that I suffered when I spoke of this, and many were my fears and persecutions. They felt so certain of my being possessed by a devil that some of them wanted to exorcize me. This did not worry me much, but I was distressed when I found my confessors unwilling to hear my confession, or when I heard that people were talking to them about me. Nevertheless, I could not be sorry that I had seen these celestial visions. I would not have exchanged a single one of them for all the blessings and delights in the world. I always regarded them as a grand mercy from the Lord, and I think they were a very great treasure. Often the Lord Himself would reassure me, and I found my love for Him growing exceedingly. I would go and complain to Him about all my trials, and I always emerged from prayer comforted and with new strength. But I did not dare to contradict my critics, for I saw that this made things worse, since they attributed my arguments to lack of humility. I discussed things with my confessor, however, and he never failed to give me great comfort if he saw that I was worried.

When the visions became more frequent, one of those who had helped me before, and who had taken my confession sometimes when the minister could not, began to say that clearly I was being deceived by the devil. He ordered me, since I had no power of resistance, always to make the sign of the Cross when I had a vision, and to snap my fingers at it, in the firm conviction that this was the devil's work. Then it would not come again. He told me to have no fear, for God would protect me and take the vision away. This command greatly distressed me, for I could not think that the vision came from anything but God. It was a terrible thing for me to do; and, as I have said, I could not possibly wish my vision to be taken from me. However, in the end I obeyed him. I prayed God frequently to free me from deception; indeed, I did so continually, with many tears, and I also invoked St Peter and St Paul. For the Lord had told me, when He first appeared to me on their festival, that they would preserve me from being deceived. I used often to see them very clearly, on my left, and that was no imaginary vision. These glorious saints were my very true lords.

The duty of snapping my fingers when I had this vision of the Lord deeply distressed me. For when I saw Him before me, I would willingly have been hacked to death rather than believe that this was of the devil. It was a heavy kind of penance for me, and so that I need not be so continually crossing myself, I used to go about with a crucifix in my hand. I carried it almost continually, but I did not snap my fingers very often, because that hurt me too much. It reminded me of the insults He had suffered from the Jews, and I begged Him to pardon me, since I was only acting out of obedience to one who was in His place, and not to blame me, seeing that he was one of the ministers whom He had Himself placed in His Church. He told me not to worry, since I was quite right to obey, and that He would Himself show them the truth. When they forbade me to pray, He seemed to me to be angry. He told me to say to them that this was tyranny. He showed me ways of making sure that these visions were not of the devil, and I will give some of them later.

Once when I was holding the cross of a rosary in my hand, He took it from me into His own; and when He returned it to me, it consisted of four large stones much more precious than diamonds—incomparably so for it is, of course, impossible to make comparisons between things seen supernaturally and the precious stones of this world; diamonds seem imperfect counterfeits beside the precious stones of a vision. On these were exquisitely incised the five wounds of Christ. He told me that henceforth this cross would appear so to me always, and so it has. I have never been able to see the wood of which it was made but only these stones. However, they have been seen by no one but myself. Once they started telling me to test my visions and resist them, these favours became much more frequent. In my efforts to divert my attention, I never ceased praying, and I seemed to be in a state of prayer even when asleep. For now my love was growing, and I would complain to the Lord, saying that I could not bear it. But desire and strive though I might to cease thinking of Him, it was beyond my power; I was as obedient as possible in every way, but I could do little or nothing about it. The Lord never released me from my obedience. But though He told me to do as I was told, He reassured me in another way by telling me how to answer my critics; and this He still does. The arguments He gave me were so strong that I felt perfectly secure.

Shortly afterwards, His Majesty began, as He had promised, to make it even plainer that it was He. There grew so great a love of God within me that I did not know who had planted it there. It was entirely supernatural; I had made no efforts to obtain it. I found myself dying of the desire to see God, and I knew no way of seeking that other life except through death. This love came to me in mighty impulses which, although less unbearable and less valuable than those that I have described before, robbed me of all power of action. Nothing gave me satisfaction, and I could not contain myself; I really felt as if my soul were being torn from me. O supreme cunning of the Lord, with what delicate skill did You work on Your miserable slave! You hid Yourself from me, and out of Your love You afflicted me with so delectable a death that my soul desired it never to cease.

No one who has not experienced these mighty impulses can possibly understand that this is no emotional unrest, nor one of those fits of uncontrollable devotion that frequently occur and seem to overwhelm the spirit. These are very low forms of prayer. Indeed, such quickenings should be checked by a gentle endeavour to become recollected, and to calm the soul. Such prayer is like the violent sobbing of children. They seem to be going to choke, but their rush of emotion is immediately checked if they are given something to drink. In the same way here, reason must step in and take command, for this may merely be a display of temperament. With reflection there comes a fear that there is some imperfection here, which may be largely physical. So the child must be quieted with a loving caress, which will draw out its love in a gentle way and not, as they say, bludgeon it. This love must flow into interior reflection, not boil over like a cooking-pot that has been put on too fierce a fire, and so spills its contents. The source of the fire must be controlled. An endeavour must be made to quench its flames with gentle tears, and not with that painful weeping that springs from the feelings I have described, and does so much damage. I used at first to shed tears of this kind which left my mind so confused and my spirit so weary that I was not fit to resume my prayers for a day or more. Great discretion is needed at first, therefore, so that everything may go on smoothly, and so that spiritual transformations may take place within. All exterior demonstrations should be carefully prevented.

The true impulses are very different. We do not pile the wood beneath the fire ourselves; it is rather as if it were already burning and we were suddenly thrown in to be consumed. The soul makes no effort to feel the pain caused it by the Lord's presence, but is pierced to the depths of its entrails, or sometimes to the heart, by an arrow, so that it does not know what is wrong or what it desires. It knows quite well that it desires God, and that the arrow seems to have been tipped with some poison which makes it so hate itself out of love of the Lord that it is willing to give up its life for Him. It is impossible to describe or explain the way in which God wounds the soul, or the very great pain He inflicts on it, so that it hardly knows what it is doing. But this is so sweet a pain that no delight in the whole world can be more pleasing. The soul, as I have said, would be glad always to be dying of this ill.

This combination of joy and sorrow so bewildered me that I could not understand how such a thing could be. O what it is to see a soul wounded! I mean one that sufficiently understands its condition as to be able to call itself wounded, and for so excellent a cause. It clearly sees that this love has come to it through no action of its own, but that out of the very great love that the Lord has for it a spark seems suddenly to have fallen on it and set it all on fire. O how often, when I am in this state, do I remember that verse of David, *As the heart panteth after the water brooks*,[1] which I seem to see literally fulfilled in myself.

When these impulses are not very strong, things appear to calm down a little, or at least the soul seeks some respite, for it does not know what to do. It performs certain penances, but hardly feels them; even if it draws blood it is no more conscious of pain than if the body were dead. It seeks ways and means to express some of its feelings for the love of God, but its initial pain is so great that I know of no physical torture that could drown it. Such medicines can bring no relief; they are on too low a level for so high a disease. But there is some alleviation and a little of the pain passes if the soul prays God to give it some remedy for its suffering, though it can see no way except death by which it can expect to enjoy its blessing complete. But there are other times when the impulses are so strong that it can do absolutely nothing. The entire body contracts; neither foot nor arm can be moved. If one is standing at the time, one falls into a sitting position as though transported, and cannot even take a breath. One only utters a few slight moans, not aloud, for that is impossible, but inwardly, out of pain.

Our Lord was pleased that I should sometimes see a vision of this kind. Beside me, on the left hand, appeared an angel in bodily form, such as I am not in the habit of seeing except very rarely. Though I often have visions of angels, I do not see them. They come to me only after the manner of the first type of vision that I described. But it was our Lord's will that I should see this angel in the following way. He was not tall but short, and very beautiful; and his face was so aflame that he appeared to be one of the highest rank of angels, who seem to be all on fire. They must be of the kind called cherubim, but they do not tell me their names. I know very well that there is a great difference between some angels and others, and between these and others still, but I could not possibly explain it. In his hands I saw a great golden spear, and at the iron tip there appeared to be a point of fire. This he plunged into my heart several times so that it penetrated to my entrails. When he pulled it out, I felt that he took them with it, and left me utterly consumed by the great love of God. The pain was so severe that it made me utter several moans. The sweetness caused by this intense pain is so extreme that one cannot possibly wish it to cease, nor is one's soul then content with anything but God. This is not a physical, but a spiritual pain, though the body has some share in it—even a considerable share. So gentle is this wooing which takes place between God and the soul that if anyone thinks I am lying, I pray God, in His goodness, to grant him some experience of it.

Throughout the days that this lasted I went about in a kind of stupor. I had no wish to look or to speak, only to embrace my pain, which was a greater bliss than all created things could give me. On several occasions when I was in this state the Lord was pleased that I should experience raptures so deep that I could not resist them even though I was not alone. Greatly to my distress, therefore, my raptures began to be talked about. Since I have had them, I have ceased to feel this pain so much, though I still feel the pain that I spoke of in a previous chapter—I do not remember which.[2] The latter is very different in many respects, and much more valuable. But when this pain of which I am now speaking begins, the Lord seems to transport the soul and throw it into an ecstasy. So there is no opportunity for it to feel its pain or suffering, for the enjoyment comes immediately. May He be blessed for ever, who has granted so many favours to one who has so ill repaid these great benefits.

CHAPTER IV[3] from *The Way of Perfection*

An exhortation to obey the Rule. Three very important matters in the spiritual life. One must strive after sublime perfection in order to accomplish so great an enterprise. How to practise prayer.

1. The greatness of the work we have undertaken. 2. Prayer. 3. The three principal aids to prayer. 4. The evils of particular friendships. 5. Special danger of these in a small community. 6. Precautions against them. 7. Mutual charity. 8. Natural and supernatural love. 9. How to regard our confessors. 10. Discretion in our intercourse with them. 11. When a second confessor is needed. 12. Precautions against worldly confessors. 13. Evils caused by unsuitable confessors.

1. You see upon how great an enterprise you have embarked for the sake of the Father Provincial, the Bishop of the diocese, and of your Order, in which all else is included, all being for the good of the Church, for which we are bound to pray as a matter of obligation. As

I said, what lives are not those bound to live who have had the courage to engage in this design, if they would not be confounded, before God and man, for their audacity? Clearly we must work hard; it is a great help to have high aspirations: by their means we may cause our actions to become great also, although there are different ways of doing so. If we endeavour to observe our Rule and Constitutions very faithfully, I hope that God will grant our petitions. I ask of you nothing new, my daughters, but only that we should keep what we have professed, as we are bound to do, although there are very diverse ways of observing it.

2. The very first chapter of our Rule bids us "Pray without ceasing"[4]. We must obey this with the greatest perfection possible, for it is our most important duty: then we shall not neglect the fasts, penances, and silence enjoined by the Rule. As you know, these are necessary if the prayer is to be genuine; prayer and self-indulgence do not go together. Prayer is the subject you have asked me to speak of: I beg of you, in return, to practise and to read, again and again, what I have already told you. Before speaking of spiritual matters, that is, of prayer. I will mention some things that must be done by those who intend to lead a life of prayer. These are so necessary that, with their help, a person who can hardly be called a contemplative may make great progress in serving God, but without them none can be a thorough contemplative: any one who imagined that she was so, would be much mistaken. May our Lord give me His grace for this task and teach me what to say that may be for His glory.[5]

3. Do not fancy, my friends and my sisters, that I am going to lay many charges on you: please God we may fulfill those that our holy Fathers enjoined and practised in our Rules and Constitutions, which include all the virtues and by performing which our predecessors earned the name of Saints. It would be an error to seek another road or to try to learn some other way. I will explain three matters only, which are in our Constitutions: it is essential for us to understand how much they help us to preserve that peace, both interior and exterior, which our Lord so strongly enjoined. The first of these is love for one another: the second, detachment from all created things: the other is true humility, which, though I mention it last, is chief of all and includes the rest.[6] The first matter, that is, mutual charity, is most important, for there is no annoyance that cannot easily be borne by those who love one another: anything must be very out of the way to cause irritation. If this commandment were observed in this world as it ought to be, I believe it would be a great help towards obeying the others, but whether we err by excess or by defect, we only succeed in keeping it imperfectly.

4. You may think there can be no harm in excessive love for one another, but no one would believe what evil and imperfections spring from this source unless they had seen it for themselves. The devil sets many snares here which are hardly detected by those who are content to serve God in a superficial way—indeed, they take such conduct for virtue—those, however, who are bent on perfection understand the evil clearly, for, little by little it deprives the will of strength to devote itself entirely to the love of God. I think this injures women even more than men, and does serious damage to the community. It prevents a nun from loving all the others equally, makes her resent any injury done to her friend, causes her to wish she had something to give her favourite and to seek for opportunities to talk to her often, and tell her how much she loves her, and other nonsense of the sort, rather than of how much she loves God. These close friendships rarely serve to forward the love of God; in fact, I believe the devil originates them so as to make factions among the religious. When a friendship has the service of God for its object, it is at once manifest that the will is not only uninfluenced by passion but it rather helps to subdue the other passions.

5. In a large convent I permit such friendships, but in St. Joseph's, where there are, and can be, no more than thirteen nuns, all must love and help one another. Keep free of partialities, for the love of God, however holy they may be, for even among brothers they are like poison. I can see no advantage in them, and matters are far worse when they exist between relatives, for then they are a perfect pest. Believe me, sisters, though I may seem to you severe in excluding these attachments, yet this promotes high perfection and quiet peace, and weak souls are spared dangerous occasions. If we are inclined to care for one person more than another (which cannot be helped, for it is but human, and we often prefer the most faulty if they more natural charm) let us control our likings firmly, and not allow ourselves to be overmastered by affections.

6. Let us love virtue and holiness and always try to prevent ourselves from being attracted by externals. O my sisters, let us not permit our will to become the slave of any save of Him Who purchased it with His Blood, or, without knowing how, we shall find ourselves caught in a trap from which we cannot escape! Lord have mercy upon us! The childish nonsense that comes from this is untold, and is so petty that no one could credit it who had not witnessed the thing. It is best not to speak of it here, lest women's foibles should be learnt by those know nothing about them, so I will give no details, although they astonish even me at times. By the mercy of God, I have never been entangled in such things myself, but perhaps this may be because I have fallen into far graver faults. However, as I have said, I have often seen it, but as I told you, in a Superior it would be ruinous. In order to guard against these partialities, great care must be taken from the very first, and this more by watchfulness and kindness than by severity. A most useful precaution is for the nuns, according to our present habit, never to be with one another nor talk together except at the appointed times, but as the Rule enjoins, for the sisters not to be together but each one alone in her cell.[7] Let there be no common workroom in St. Joseph's, for although this is a praiseworthy custom, silence is better kept when one is alone. Solitude is very helpful to persons who practise prayer, and since prayer is the mortar which keeps this house together, we must learn to like what promotes it.[8]

7. To return to speak of our charity for one another. It seems superfluous to insist on this, for who would be so boorish as not to love those with whom they associate and live, cut off as they are from all conversation, intercourse and recreation with any one outside the house, whilst believing that they bear a mutual love for God, as He has for all of them, since for His sake they have left everything? More especially as goodness always attracts love, and, by the blessing of God, I trust that the nuns of this convent will always be good. Therefore, I do not think there is much need for me to persuade you to love each other. But as regards the nature of this love and of the virtuous love that I wish you all to feel, and the means of knowing whether we possess this greatest of virtues—for it must be a very great virtue since our Lord so often enjoins it on us, as He did most stringently upon His apostles—of this I will speak to you for a short time as well as my inaptitude will allow. If you find the matter explained in any other books, you need not read mine, for I very frequently do not understand what I am talking about, unless our Lord enlightens me.[9]

8. I intend treating of two kinds of love: one which is entirely spiritual, free from any sort of affection or natural tenderness which could tarnish its purity, and another which is spiritual but mingled with the frailty and weakness of human nature. The latter is good and seems lawful, being such as is felt between relatives and friends, and is that which I have mentioned before. The first of these two ways of loving, and the one that I will discuss, is unmixed with any kind of passion that would disturb its harmony. This love, exercised with moderation and discretion, is profitable in every way, particularly when borne towards holy people or confessors, for that which seems only natural is then changed into virtue.[10] At times, however, these two kinds of love seem so combined that it is difficult to distinguish them from one another, especially as regards a confessor. When persons who practice prayer discover that their confessor is a holy man who understands their spiritual state, they feel a strong affection for him; the devil then opens a perfect battery of scruples on the soul, which, as he intends, greatly disturb it, especially if the priest is leading his penitent to higher perfection. Then the evil one torments his victim to such a pitch that she leaves her director, so that the temptation gives her no peace either in one way or the other.

9. In such a case it is best not to think about whether you like you confessor or not, or whether you wish to like him. If we feel friendship for those who benefit our bodies, why should we not feel as great a friendship for those who strive and labour to benefit our souls? On the contrary, I think a liking for my confessor is a great help to my progress if he is holy and spiritual, and if I see that he endeavours to profit my soul. Human nature is so weak that this feeling is often a help to our undertaking great things in God's service.

10. If, however, the confessor be a man of indifferent character, we must not let him know of our liking for him. Great prudence and caution are necessary on account of the difficulty of knowing his disposition: it is best, on this account, to conceal our feelings from him. You should believe that your friendship for him is harmless and think no more about it. You may follow this advice when you see that all your confessor says tends to profit your soul and when you discover no levity in him, but are conscious that he lives in the fear of God: any one can detect this at once unless she wilfully blinds herself. If this be so, do not allow any temptation to trouble you about your liking for him— despise it, think no more about it, and the devil will grow tired and leave you alone. If, however, the confessor appears worldly-minded, be most guarded in every way; do not talk with him even when he converses on religious subjects, but make your confession briefly and say no more. It would be best to tell the Prioress that he does not suit your soul and to ask for someone else; this is the wisest course to take if it is possible, and can be done without injuring his reputation. I trust in God that it may be feasible for you.

11. In these and other difficulties by which the devil may seek to ensnare us, it would be best, when you are doubtful as to what course to pursue, for you to consult some learned person, as the nuns are permitted to do,[11] to make your confession to him, and to follow his advice in the case, lest some great mistake should be made in remedying the evil. How many people go astray in the world for want of seeking guidance, especially in what affects their neighbours' interests! Some redress must be sought, for when the devil starts such works, unless he is stopped at once, the matter will become serious, therefore my advice about changing confessors is the best, and I trust in God that you will be able to do so.

12. Be convinced of the importance of this: the thing is dangerous, a hell in itself, and injurious to every one. Do not wait until much harm has come of it, but stop the matter at once in every feasible way: this may be done with a clear conscience. I trust, however, that God will prevent those vowed to a life of prayer from becoming attached to any one who does not serve God fervently, as He certainly will, unless they omit to practise prayer and to strive after perfection, as we profess to do in this house. If the nuns see that the confessor does not understand their language nor cares to speak of God they cannot like him, for he differs from them. If he is of such a character, he will have extremely few chances of doing any harm here, and unless he is very foolish he will neither trouble himself about the servants of God, nor disturb those who have few pleasures and little or no opportunity of following their own way.

13. Since I have begun speaking on this subject, I may say that this is the only harm, or at any rate the greatest harm, that the devil can do within enclosed convents. It takes long to discover, so that great damage may have been wrought to perfection without any one knowing how, for if the confessor is worldly himself, he will treat the defect lightly in others. Deliver us, O Lord, for Thine own sake, from such misfortunes!

It is enough to unsettle all the nuns if their conscience tells them one thing and their confessor another. Where they are allowed no other director I do not know what to do, nor how to quiet their minds, for he who ought to bring them peace and counsel is the very author of the evil. There must be a great deal of trouble, resulting in much harm, in some places from these misplaced partialities, as I have seen in certain convents to my great sorrow, therefore you need not be surprised at the pains I have taken to make you understand the danger.

End notes

1 *L'Etre et le Neant* (Paris: Gallimand, 1943), translated by Hazel E. Barnes (New York: Philosophical Library, 1956).

2 'Meaning'. *Philosophical Review*, lxvi, no. 3 (July, 1957). 377-88.

3 Sec *Romans*, VII, 23; and the *Confessions*, bk vIII, pt V.

26 *Civilization and its Discontents*

Sigmund Freud

Before we go on to enquire from what quarter an interference might arise, this recognition of love as one of the foundations of civilization may serve as an excuse for a digression which will enable us to fill in a gap which we left in an earlier discussion. We said there that man's discovery that sexual (genital) love afforded him the strongest experiences of satisfaction, and in fact provided him with the prototype of all happiness, must have suggested to him that he should continue to seek the satisfaction of happiness in his life along the path of sexual relations and that he should make genital erotism the central point of his life. We went on to say that in doing so he made himself dependent in a most dangerous way on a portion of the external world, namely, his chosen love-object, and exposed himself to extreme suffering if he should be rejected by that object or should lose it through unfaithfulness or death. For that reason the wise men of every age have warned us most emphatically against this way of life; but in spite of this it has not lost its attraction for a great number of people.

A small minority are enabled by their constitution to find happiness, in spite of everything, along the path of love. But far-reaching mental changes in the function of love are necessary before this can happen. These people make themselves independent of their object's acquiescence by displacing what they mainly value from being loved on to loving; they protect themselves against the loss of the object by directing their love, not to single objects but to all men alike; and they avoid the uncertainties and disappointments of genital love by turning away from its sexual aims and transforming the instinct into an impulse with an inhibited aim. What they bring about in themselves in this way is a state of evenly suspended, steadfast, affectionate feeling, which has little external resemblance any more to the stormy agitations of genital love, from which it is nevertheless derived. Perhaps St. Francis of Assisi went furthest in thus exploiting love for the benefit of an inner feeling of happiness. Moreover, what we have recognized as one of the techniques for fulfilling the pleasure principle has often been brought into connection with religion; this connection may lie in the remote regions where the distinction between the ego and objects or between objects themselves is neglected. According to one ethical view, whose deeper motivation will become clear to us presently,[1] this readiness for a universal love of mankind and the world represents the highest standpoint which man can reach. Even at this early stage of the discussion I should like to bring forward my two main objections to this view. A love that does not discriminate seems to me to forfeit a part of its own value, by doing an injustice to its object; and secondly, not all men are worthy of love.

V

Psychoanalytic work has shown us that it is precisely these frustrations of sexual life which people known as neurotics cannot tolerate. The neurotic creates substitutive satisfactions for himself in his symptoms, and these either cause him suffering in themselves or become sources of suffering for him by raising difficulties in his relations with his environment and the society he belongs to. The latter fact is easy to understand; the former presents us with a new problem. But civilization demands other sacrifices besides that of sexual satisfaction.

We have treated the difficulty of cultural development as a general difficulty of development by tracing it to the inertia of the libido, to its disinclination to give up an old position for a new one.[1] We are saying much the same thing when we derive the antithesis between civilization and sexuality from the circumstance that sexual love is a relationship between two individuals in which a third can only be superfluous or disturbing, whereas civilization depends on relationships between a considerable number of individuals. When a love-relationship is at its height there is no room left for any interest in the environment; a pair of lovers are sufficient to themselves, and do not even need the child they have in common to make them happy. In no other case does Eros so clearly betray the core of his being, his purpose of making one out of more than one; but when he has achieved this in the proverbial way through the love of two human beings, he refuses to go further.

So far, we can quite well imagine a cultural community consisting of double individuals like this, who, libidinally satisfied in themselves, are connected with one another through the bonds of common work and common interests. If this were so, civilization would not have to withdraw any energy from sexuality. But this desirable state of things does not, and never did, exist. Reality shows us that civilization is not content with the ties we have so far allowed it. It aims at binding the members of the community together in a libidinal way as well and employs every means to that end. It favours every path by which strong identifications can be established between the members of the community, and it summons up aim-inhibited libido on the largest scale so as to strengthen the communal bond by relations of friendship. In order for these aims to be fulfilled, a restriction upon sexual life is unavoidable. But we are unable to understand what the necessity is which forces civilization along this path and which causes its antagonism to sexuality. There must be some disturbing factor which we have not yet discovered.

The clue may be supplied by one of the ideal demands, as we have called them,[2] of civilized society. It runs: "Thou shalt love thy neighbour as thyself.' It is known throughout the world and is undoubtedly older than Christianity, which puts it forward as its proudest claim. Yet it is certainly not very old; even in historical times it was still strange to mankind. Let us adopt a naive attitude towards it, as though we were hearing it for the first time; we shall be unable then to suppress a feeling of surprise and bewilderment. Why should we do it? What good will it do us? But, above all, how shall we achieve it? How can it be possible? My love is something valuable to me which I ought not to throw away without reflection. It imposes duties on me for whose fulfilment I must be ready to make sacrifices. If I love someone, he must deserve it in some way. (I leave out of account the use he may be to me, and also his possible significance for me as a sexual object, for neither of these two kinds of relationship comes into question where the precept to love my neighbour is concerned.) He deserves it if he is so like me in important ways that I can love myself in him; and he deserves it if he is so much more perfect than myself that I can love my ideal of my own self in him. Again, I have to love him if he is my friend's son, since the pain my friend would feel if any harm came to him would be my pain too—I should have to share it. But if he is a stranger to me and if he cannot attract me by any worth of his own or any significance that he may already have acquired for my emotional life, it will be hard for me to love him. Indeed, I should be wrong to do so, for my love is valued by all my own people as a sign of my preferring them, and it is an injustice to them if I put a stranger on a par with them. But if I am to love him (with this universal love) merely because he, too, is an inhabitant of this earth, like an insect, an earth-worm or a grass-snake, then I fear that only a small modicum of my love will fall to his share—not by any possibility as much as, by the judgement of my reason, I am entitled to retain for myself. What is the point of a precept enunciated with so much solemnity if its fulfilment cannot be recommended as reasonable?

On closer inspection, I find still further difficulties. Not merely is this stranger in general unworthy of my love; I must honestly confess that he has more claim to my hostility and even my hatred. He seems not to have the least trace of love for me and shows me not the slightest consideration. If it will do him any good he has no hesitation in injuring me, nor does he ask himself whether the amount of advantage he gains bears any proportion to the extent of the harm he does to me. Indeed, he need not even obtain an advantage; if he can satisfy any sort of desire by it, he thinks nothing of jeering at me, insulting me, slandering me and showing his superior power; and the more secure he feels and the more helpless I am, the more certainly I can expect him to behave like this to me. If he behaves differently, if he shows me consideration and forbearance as a stranger, I am ready to treat him in the same way, in any case and quite apart from any precept. Indeed, if this grandiose commandment had run 'Love thy neighbour as thy neighbour loves thee', I should not take exception to it. And there is a second commandment, which seems to me even more incomprehensible and arouses still stronger opposition in me. It is 'Love thine enemies'. If I think it over, however, I see that I am wrong in treating it as a greater imposition. At bottom it is the same thing.[3]

I think I can now hear a dignified voice admonishing me: 'It is precisely because your neighbour is not worthy of love, and is on the contrary your enemy, that you should love him as yourself.' I then understand that the case is one like that of Credo quia absurdum.[4]

Now it is very probable that my neighbour, when he is enjoined to love me as himself, will answer exactly as I have done and will repel me for the same reasons. I hope he will not have the same objective grounds for doing so, but he will have the same idea as I have. Even so, the behaviour of human beings shows differences, which ethics, disregarding the fact that such differences are determined, classifies as 'good' or 'bad'. So long as these undeniable differences have not been removed, obedience to high ethical demands entails damage to the aims of civilization, for it puts a positive premium on being bad. One is irresistibly reminded of an incident in the French Chamber when capital punishment was being debated. A member had been passionately supporting its abolition and his speech was being received with tumultuous applause, when a voice from the hall called out: 'Que messieurs les assassins commencent!'[5]

The element of truth behind all this, which people are so ready to disavow, is that men are not gentle creatures who want to be loved, and who at the most can defend themselves if they are attacked; they are, on the contrary, creatures among whose instinctual endowments is to be reckoned a powerful share of aggressiveness. As a result, their neighbour is for them not only a potential helper or sexual object, but also someone who tempts them to satisfy their aggressiveness on him, to exploit his capacity for work without compensation, to use him sexually without his consent, to seize his possessions, to humiliate him, to cause him pain, to torture and to kill him. Homo homini lupus.[6] Who, in the face of all his experience of life and of history, will have the courage to dispute this assertion? As a rule this cruel aggressiveness waits for some provocation or puts itself at the service of some other purpose, whose goal might also have been reached by milder measures. In circumstances that are favourable to it, when the mental counter-forces which ordinarily inhibit it are out of action, it also manifests itself spontaneously and reveals man as a savage beast to whom consideration towards his own kind is something alien. Anyone who calls to mind the atrocities committed during the racial migrations or the invasions of the Huns, or by the people known as Mongols under Jenghiz Khan and Tamerlane, or at the capture of Jerusalem by the pious Crusaders, or even, indeed, the horrors of the recent World War—anyone who calls these things to mind will have to bow humbly before the truth of this view.

The existence of this inclination to aggression, which we can detect in ourselves and justly assume to be present in others, is the factor which disturbs our relations with our neighbour and which forces civilization into such a high expenditure [of energy]. In consequence of this primary mutual hostility of human beings, civilized society is perpetually threatened with disintegration. The interest of work in common would not hold it together; instinctual passions are stronger than reasonable interests. Civilization has to use its utmost efforts in order to set limits to man's aggressive instincts and to hold the manifestations of them in check by psychical reaction-formations. Hence, therefore, the use of methods intended to incite people into identifications and aim-inhibited relationships of love, hence the restriction upon sexual life, and hence too the ideal's commandment to love one's neighbour as oneself—a commandment which is really justified by the fact that nothing else runs so strongly counter to the original nature of man. In spite of every effort, these endeavours of civilization have not so far achieved very much. It hopes to prevent the crudest excesses of brutal violence by itself assuming the right to use violence against criminals, but the law is not able to lay hold of the more cautious and refined manifestations of human aggressiveness.

The time comes when each one of us has to give up as illusions the expectations which, in his youth, he pinned upon his fellow men, and when he may learn how much difficulty and pain has been added to his life by their ill-will. At the same time, it would be unfair to reproach civilization with trying to eliminate strife and competition from human activity. These things are undoubtedly indispensable. But opposition is not necessarily enmity; it is merely misused and made an occasion for enmity.

The communists believe that they have found the path to deliverance from our evils. According to them, man is wholly good and is well-disposed to his neighbour; but the institution of private property has corrupted his nature. The ownership of private wealth gives the individual power, and with it the temptation to ill-treat his neighbour; while the man who is excluded from possession is bound to rebel in hostility against his oppressor. If private property were abolished, all wealth held in common, and everyone allowed to share in the enjoyment of it, ill-will and hostility would disappear among men. Since everyone's needs would be satisfied, no one would have any reason to regard another as his enemy; all would willingly undertake the work that was necessary. I have no concern with any economic criticisms of the communist system; I cannot enquire into whether the abolition of private property is expedient or advantageous.[7] But I am able to recognize that the psychological premises on which the system is based are an untenable illusion. In abolishing private property we deprive the human love of aggression of one of its instruments, certainly a strong one, though certainly not the strongest; but we have in no way altered the differences in power and influence which are misused by aggressiveness, nor have we altered anything in its nature. Aggressiveness was not created by property. It reigned almost without limit in primitive times, when property was still very scanty, and it already shows itself in the nursery almost before property has given up its primal, anal form; it forms the basis of every relation of affection and love among people (with the single exception, perhaps, of the mother's relation to her male child[8]). If we do away with personal rights over material wealth, there still remains prerogative in the field of sexual relationships, which is bound to become the source of the strongest dislike and the most violent hostility among men who in other respects are on an equal footing. If we were to remove this factor, too, by allowing complete freedom of sexual life and thus abolishing the family, the germ-cell of civilization, we cannot, it is true, easily foresee what new paths the development of civilization could take; but one thing we can expect, and that is that this indestructible feature of human nature will follow it there.

It is clearly not easy for men to give up the satisfaction of this inclination to aggression. They do not feel comfortable without it. The advantage which a comparatively small cultural group offers of allowing this instinct an outlet in the form of hostility against intruders is not to be despised. It is always possible to bind together a considerable number of people in love, so long as there are other people left over to receive the manifestations of their aggressiveness. I once discussed the phenomenon that it is precisely communities with adjoining territories, and related to each other in other ways as well, who are engaged in constant feuds and in ridiculing each other— like the Spaniards and Portuguese, for instance, the North Germans and South Germans, the English and Scotch, and so on.[9] I gave this phenomenon the name of 'the narcissism of minor differences', a name which does not do much to explain it. We can now see that it is a convenient and relatively harmless satisfaction of the inclination to aggression, by means of which cohesion between the members of the community is made easier. In this respect the Jewish people, scattered everywhere, have rendered most useful services to the civilizations of the countries that have been their hosts; but unfortunately all the massacres of the Jews in the Middle Ages did not suffice to make that period more peaceful and secure for their Christian fellows. When once the Apostle Paul had posited universal love between men as the foundation of his Christian community, extreme intolerance on the part of Christendom towards those who remained outside it became the inevitable consequence. To the Romans, who had not founded their communal life as a State upon love, religious intolerance was something foreign, although with them religion was a concern of the State and the State was permeated by religion. Neither was it an unaccountable chance that the dream of a Germanic world-dominion called for anti-semitism as its complement; and it is intelligible that the attempt to establish a new, communist civilization in Russia should find its psychological support in the persecution of the bourgeois. One only wonders, with concern, what the Soviets will do after they have wiped out their bourgeois.

If civilization imposes such great sacrifices not only on man's sexuality but on his aggressivity, we can understand better why it is hard for him to be happy in that civilization. In fact, primitive man was better off in knowing no restrictions of instinct. To counterbalance this, his prospects of enjoying this happiness for any length of time were very slender. Civilized man has exchanged a portion of his possibilities of happiness for a portion of security. We must not forget, however, that in the primal family only the head of it enjoyed this instinctual freedom; the rest lived in slavish suppression. In that primal period of civilization, the contrast between a minority who enjoyed the advantages of civilization and a majority who were robbed of those advantages was, therefore, carried to extremes. As regards the primitive peoples who exist today, careful researches have shown that their instinctual life is by no means to be envied for its freedom. It is subject to restrictions of a different kind but perhaps of greater severity than those attaching to modern civilized man.

When we justly find fault with the present state of our civilization for so inadequately fulfilling our demands for a plan of life that shall make us happy, and for allowing the existence of so much suffering which could probably be avoided—when, with unsparing criticism, we try to uncover the roots of its imperfection, we are undoubtedly exercising a proper right and are not showing ourselves enemies of civilization. We may expect gradually to carry through such alterations in our civilization as will better satisfy our needs and will escape our criticisms. But perhaps we may also familiarize ourselves with the idea that there are difficulties attaching to the nature of civilization which will not yield to any attempt at reform. Over and above the tasks of restricting the instincts, which we are prepared for, there forces itself on our notice the danger of a state of things which might be termed 'the psychological poverty of groups'.[10] This danger is most threatening where the bonds of a society are chiefly constituted by the identification of its members with one another, while individuals of the leader type do not acquire the importance that should fall to them in the formation of a group.[11] The present cultural state of America would give us a good opportunity for studying the damage to civilization which is thus to be feared. But I shall avoid the temptation of entering upon a critique of American civilization; I do not wish to give an impression of wanting myself to employ American methods.

End Notes

1 [For some remarks on Freud's use of the concept of 'psychical inertia' in general, see an Editor's footnote to Freud, 1915f, *Standard Ed.*, 14, 272.]

2 ['"Civilized" Sexual Morality' (1908d), *Standard Ed.*, 9, 199.]

3 A great imaginative writer may permit himself to give expression— jokingly, at all events—to psychological truths that are severely proscribed. Thus Heine confesses: 'Mine is a most peaceable disposition. My wishes are: a humble cottage with a thatched roof, but a good bed, good food, the freshest milk and butter, flowers before my window, and a few fine trees before my door; and if God wants to make my happiness complete, he will grant me the joy of seeing some six or seven of my enemies hanging from those trees. Before their death I shall, moved in my heart, forgive them all the wrong they did me in their lifetime. One must, it is true, forgive one's enemies—but not before they have been hanged.'

4 [See Chapter V of *The Future of an Illusion* (1927c). Freud returns to the question of the commandment to love one's neighbour as oneself below, on p. 89f.].

5 ['It's the murderers who should make the first move.']

6 ['Man is a wolf to man.' Derived from Plautus, *Asinaria* II, iv, 88.]

7 Anyone who has tasted the miseries of poverty in his own youth and has experienced the indifference and arrogance of the well-to-do, should be safe from the suspicion of having no understanding or good will towards endeavours to fight against the inequality of wealth among men and all that it leads to. To be sure, if an attempt is made to base this fight upon an abstract demand, in the name of justice, for equality for all men, there is a very obvious objection to be made—that nature, by endowing individuals with extremely unequal physical attributes and mental capacities, has introduced injustices against which there is no remedy.

8 [Cf. a footnote to Chapter VI of *Group Psychology* (1921c), Standard Ed., 18, 101n. A rather longer discussion of the point occurs near the end of Lecture XXXIII of the *New Introductory Lectures* (1933a).]

9 [See Chapter VI of *Group Psychology* (192k), Standard Ed., 18, 101, and 'The Taboo of Virginity' (1918a), ibid., 11,199.]

10 [The German '*psychologisches Elend*' seems to be a version of Janet's expression '*misère psyckologique*' applied by him to describe the incapacity for mental synthesis which he attributes to neurotics.]

11 Cf. *Group Psychology and the Analysis of the Ego* (1921c).

27 *The Hebrew Bible*: Genesis

IN THE BEGINNING God created[a] the heavens and the earth. [2]The earth was without form and void, and darkness was upon the face of the deep; and the Spirit[b] of God was moving over the face of the waters.

[3]And God said, "Let there be light"; and there was light. [4]And God saw that the light was good; and God separated the light from the darkness. [5]God called the light Day, and the darkness he called Night. And there was evening and there was morning, one day.

[6]And God said, "Let there be a firmament in the midst of the waters, and let it separate the waters from the waters." [7]And God made the firmament and separated the waters which were under the firmament from the waters which were above the firmament. And it was so. [8]And God called the firmament Heaven. And there was evening and there was morning, a second day.

[9]And God said, "Let the waters under the heavens be gathered together into one place, and let the dry land appear." And it was so. [10]God called the dry land Earth, and the waters that were gathered together he called Seas. And God saw that it was good. [11]And God said, "Let the earth put forth vegetation, plants yielding seed, and fruit trees bearing fruit in which is their seed, each according to its kind, upon the earth." And it was so. [12]The earth brought forth vegetation, plants yielding seed according to their own kinds, and trees bearing fruit in which is their seed, each according to its kind. And God saw that it was good. [13]And there was evening and there was morning, a third day.

[14]And God said, "Let there be lights in the firmament of the heavens to separate the day from the night; and let them be for signs and for seasons and for days and years, [15]and let them be lights in the firmament of the heavens to give light upon the earth." And it was so. [16]And God made the two great lights, the greater light to rule the day, and the lesser light to rule the night; he made the stars also. [17]And God set them in the firmament of the heavens

to give light upon the earth, [18]to rule over the day and over the night, and to separate the light from the darkness. And God saw that it was good. [19]And there was evening and there was morning, a fourth day.

[20]And God said, "Let the waters bring forth swarms of living creatures, and let birds fly above the earth across the firmament of the heavens." [21]So God created the great sea monsters and every living creature that moves with which the waters swarm according to their kinds, and every winged bird according to its kind. And God saw that it was good. [22]And God blessed them, saying, "Be fruitful and multiply and fill the waters in the seas, and let birds multiply on the earth."[23]And there was evening and there was morning, a fifth day.

[24]And God said, "Let the earth bring forth living creatures according to their kinds: cattle and creeping things and beasts of the earth according to their kinds." And it was so. [25]And God made the beasts of the earth according to their kinds and the cattle according to their kinds, and everything that creeps upon the ground according to its kind. And God saw that it was good.

[26]Then God said, "Let us make man in our image, after our likeness; and let them have dominion over the fish of the sea, and over the birds of the air, and over the cattle, and over all the earth, and over every creeping thing that creeps upon the earth." [27]So God created man in his own image, in the image of God he created him; male and female he created them. [28]And God blessed them, and God said to them, "Be fruitful and multiply, and fill the earth and subdue it; and have dominion over the fish of the sea and over the birds of the air and over every living thing that moves upon the earth." [29]And God said, "Behold, I have given you every plant yielding seed which is upon the face of all the earth, and every tree with seed in its fruit; you shall have them for food. [30]And to every beast of the earth, and to every bird of the air, and to everything that creeps on the earth, everything that has the breath of life, I have given every green plant for food." And it was so. [31]And God saw everything that he had made, and behold, it was very good. And there was evening and there was morning, a sixth day.

[2]Thus the heavens and the earth were finished, and all the host of them. [2]And on the seventh day God finished his work which he had done, and he rested on the seventh day from all his work which he had done. [3]So God blessed the seventh day and hallowed it, because on it God rested from all his work which he had done in creation.

[4]These are the generations of the heavens and the earth when they were created.

In the day that the Lord God made the earth and the heavens, [5]when no plant of the field was yet in the earth and no herb of the field had yet sprung up - for the Lord God had not caused it to rain upon the earth, and there was no man to till the ground; [6]but a mist[c] went up from the earth and watered the whole face of the ground - [7]then the Lord God formed man of dust from the ground, and breathed into his nostrils the breath of life; and man became a living being. [8]And the Lord God planted a garden in Eden, in the east; and there he put the man whom he had formed. [9]And out of the ground the Lord God made to grow every tree that is pleasant to the sight and good for food, the tree of life also in the midst of the garden, and the tree of the knowledge of good and evil.

A river flowed out of Eden to water the garden, and there it divided and became four rivers. [11]The name of the first is Pishon; it is the one which flows around the whole land of Hav'ilah, where there is gold; [12]and the gold of that land is good; bdellium and onyx stone are there. [13]The name of the second river is Gihon; it is the one which flows around the whole land of Cush. [14]And the name of the third river is Tigris, which flows east of Assyria. And the fourth river is the Euphra'tes.

[15]The Lord God took the man and put him in the garden of Eden to till it and keep it. [16]And the Lord God commanded the man, saying, "You may freely eat of every tree of the garden; [17]but of the tree of the knowledge of good and evil you shall not eat, for in the day that you eat of it you shall die."

[18]Then the Lord God said, "It is not good that the man should be alone; I will make him a helper fit for him." [19]So out of the ground the Lord God formed every beast of the field and every bird of the air, and brought them to the man to see what he would call them; and whatever the man called every living creature, that was its name. [20]The man gave names to all cattle, and to the birds of the air, and to every beast of the field; but for the man there was not found a helper fit for him. [21]So the Lord God caused a deep sleep to fall upon the man, and while he slept took one of his ribs and closed up its place with flesh; [22]and the rib which the Lord God had taken from the man he made into a woman and brought her to the man. [23]Then the man said,

> "This at last is bone of my bones and flesh of my flesh; she shall be called Woman,[d]
> because she was taken out of Man."[e]

[24]Therefore a man leaves his father and his mother and cleaves to his wife, and they become one flesh. [25]And the man and his wife were both naked, and were not ashamed.

[3]Now the serpent was more subtle than any other wild creature that the Lord God had made. He said to the woman, "Did God say, 'You shall not eat of any tree of the garden'?" [2]And the woman said to the serpent, "We may eat of the fruit of the trees of the garden; [3]but God said, 'You shall not eat of the fruit of the tree which is in the midst of the garden, neither shall you touch it, lest you die.'" [4]But the serpent said to the woman, "You will not die. [5]For God knows that when you eat of it your eyes will be opened, and you will be like God, knowing good and evil." [6]So when the woman saw that the tree was good for food, and that it was a delight to the eyes, and that the tree was to be desired to make one wise, she took of its fruit and ate; and she also gave some to her husband, and he ate. [7]Then the eyes of both were opened, and they knew that they were naked; and they sewed fig leaves together and made themselves aprons.

[8]And they heard the sound of the Lord God walking in the garden in the cool of the day, and the man and his wife hid themselves from the presence of the Lord God among the trees of the garden. [9]But the Lord God called to the man, and said to him, "Where are you?" [10]And he said, "I heard the sound of thee in the garden, and I was afraid, because I was naked; and I hid myself." [11]He said, "Who told you that you were naked? Have you eaten of the tree of which I commanded you not to eat?" [12]The man said, "The woman whom thou gavest to be with me, she gave me fruit of the tree, and I ate." [13]Then the Lord God said to

the woman, "What is this that you have done?" The woman said, "The serpent beguiled me, and I ate."

[14]The Lord God said to the serpent,
"Because you have done this,
cursed are you above all cattle,
and above all wild animals;
upon your belly you shall go,
and dust you shall eat
all the days of your life.
[15]I will put enmity between you and
the woman,
and between your seed and her
seed;
he shall bruise your head,
and you shall bruise his heel."
[16]To the woman he said,
"I will greatly multiply your pain in
childbearing;
in pain you shall bring forth
children,
yet your desire shall be for your
husband,
and he shall rule over you."
[17]And to Adam he said,
"Because you have listened to the
voice of your wife,
and have eaten of the tree
of which I commanded you,
'You shall not eat of it,'
cursed is the ground because of you;
in toil you shall eat of it all the
days of your life;
[18]thorns and thistles it shall bring
forth to you;
and you shall eat the plants of the field.
[19]In the sweat of your face
you shall eat bread
till you return to the ground,
for out of it you were taken;
you are dust,
and to dust you shall return."

[20]The man called his wife's name Eve, because she was the mother of all living. [21]And the

Lord God made for Adam and for his wife garments of skins, and clothed them.

[22]Then the Lord God said, "Behold, the man has become like one of us, knowing good and evil; and now, lest he put forth his hand and take also of the tree of life, and eat, and live forever;—[23]therefore the Lord God sent him forth from the garden of Eden, to till the ground from which he was taken. [24]He drove out the man; and at the east of the garden of Eden he placed the cherubim, and a flaming sword which turned every way, to guard the way to the tree of life.

[4]Now Adam knew Eve his wife, and she conceived and bore Cain, saying, "I have gotten[g] a man with the help of the Lord." [2]And again, she bore his brother Abel. Now Abel was a keeper of sheep, and Cain a tiller of the ground. [3]In the course of time Cain brought the Lord an offering of the fruit of the ground, [4]and Abel brought of the firstlings of his flock and of the fat portions. And the Lord had regard for Abel and his offering,...

End notes

a Or *When God began to create*

b Or *wind*

c Or *flood*

d Heb *ishshah*

e Heb *ish*

f The name in Hebrew resembles the word for *living*

g Heb *qanah*, get

1.1-2.4a: The Priestly story of creation. Out of original chaps God created an orderly world in which he assigned a preeminent place to man. **1:** Probably a preface to the whole story, though possibly introductory to v. 3: *When God began to create* (note *a*) ... *God said* (compare 2.4b-7). The ancients believed the world originated from and was founded upon a watery chaos (*the deep*; compare Ps.24.1,2), portrayed as a dragon in various myths (Is.51.9). **3-5:** Creation by the word of God (Ps.33.6-9) expresses God's absolute lordship and prepares for the doctrine of creation out of nothing (2 Macc.7.28). Light was created first (2 Cor.4.6), even before the sun, and was *separated* from *night*, a remnant of uncreated darkness (v. 2). Since the Jewish day began with sundown, the order is *evening* and *morning*. **6-8:** A *firmament*, or solid dome (Job 37.18), separated the upper from the lower waters (Ex.20.4; Ps. 148.4). See 7.11 n. **9-10:** The *seas*, a portion of the watery chaos, were assigned boundaries at the edge of the earth (Ps.139.9; Pr.8.29), where they continue to menace God's creation (Jer.5.22; Ps. 104.7-9). **11-13:** *Vegetation* was created only indirectly by God; his creative command was directed to *the earth*. **14-19:** The sun, moon, and stars are not divine powers that control man's destiny, as was believed in antiquity, but are only *lights*. Implicitly worship of the heavenly host is forbidden (Dt.4.19; Zeph.1.5). **20-23:** The creation of birds and fishes. *Sea monsters*, see Pss.74.13; 104.25-26. **24-25:** God's command for the earth to *bring forth* (compare v. 11) suggests that the animals are immediately bound to *the ground* and only indirectly related to God, in contrast with man. **26-27:** The solemn divine decision emphasizes man's supreme place at the climax of God's creative work. **26:** The plural *us, our* (3.22; 11.7; Is.6.8) probably refers to the divine beings who compose God's heavenly court (1 Kg.22.19; Job 1.6). Made in *the image of God*, man is the creature through whom God manifests his rule on earth. The language reflects "royal theology" in which, as in Egypt, the king was the "image of God." **27:** *Him, them:* man was not created to be alone but is *male and female* (2.18-24). *Man*, the Hebrew word is "adam," a collective, referring to mankind. 28: As God's representative, man is given *dominion* (Ps.8.6-8). **29-30:** His dominion is limited, as shown by the vegetarian requirement, modified in Noah's time (9.2-3); it is to be benevolent and peaceful (compare Is.11.6-8). **31:** *Very good* (vv. 4,10,12. etc.), corresponding perfectly to God's purpose. 2.1-3: The verb rested (Hebrew "*shabat*") is the basis of the noun sabbath (Ex.31.12-17).

2.4b-3.24: The creation and the fall of man. This is a different tradition from that in 1.1-2.4a, as evidenced by the flowing style and the different order of events, e.g. man is created before vegetation, animals, and woman. **6:** *A mist* (or *flood*) probably refers to the water which surged up from the subterranean ocean, the source of fertility (49.25). **7:** The word-play on *man* ('adham) and *ground* ('adhamah) introduces a motif characteristic of this early tradition: man's relation to the ground from which he was *formed*, like a potter molds clay (Jer.18.6). Man is not body and soul (a Greek distinction) but is dust animated by the Lord God's *breath* or "spirit" which constitutes him a *living being* or psycho-physical self (Ps. 104.29-30; Job 34.14-15). **8-9:** *Eden*, meaning "delight," is a "garden of God" (Is.51.3; Ezek.31.8-9; Jl.2.3) or divine park. **9:** The *tree of life* was believed to confer eternal life (3.22; see Pr.3.18 n.; Rev.22.2,14,19), as the *tree of the knowledge of good and evil* confers wisdom (see 2 Sam.l4.17;Is.7.15). **10-14:** The rivers, springing from the subterranean ocean (v. 6), flowed out to the four corners of the known historical world. **15-17:** Man is given a task: to *till* and *keep* the garden. The prohibition against eating the forbidden fruit (3.3) stresses God's lordship and man's obedience. **18:** *To be alone* is not good, for man is social by nature (see 1.27 n.). *A helper fit for him* means a partner who is suitable for him, who completes his being. **19:** Naming the animals signifies man's dominion over them (compare 1.28). **21-23:** The deep affinity between man and woman is portrayed in the statement that God made the woman from the man's *rib*. **24-25:** Sex is not regarded as evil but as a God-given impulse which draws man and woman together so that *they become one flesh*. **25:** The two were unashamedly *naked*, a symbol of their guiltless relation to God and to one another. **3.1-7:** The temptation begins with the insinuation of doubt (vv. 1-3), increases as suspicion is cast upon God's motive (vv. 4-5), and becomes irresistible when the couple sense the possibilities of freedom (v. 6). **I:** *The serpent*, one of the wild creatures, distinguished by uncanny wisdom (Mt.10.16); there is a hint of a seductive power in man's environment, hostile to God. **5:** *Like God*: perhaps "like gods" (Septuagint), the divine beings of the heavenly court (v. 22; 1.26 n.). *Knowing good and evil*, see 2.9 n. **7:** Bodily shame (2.25) symbolizes anxiety about broken relationship with God. **8-13:** Anxiety leads to a guilty attempt to hide from God (Ps.139.7-12), described anthropomorphically as strolling in his garden. **14-15:** The Curse contains an old explanation of why the serpent crawls rather than walks and why men are instinctively hostile to it. **16:** This divine judgement contains an old explanation of woman's pain in childbirth, her sexual *desire* for her husband (i.e. her motherly impulse, compare 30.1), and her subordinate position to man in ancient society. **17-19:** An explanation of man's struggle to eke an existence from the soil. Work is not essentially evil (2.15) but it becomes *toil* as a result of man's broken relationship with his Creator. **17:** The Hebrew word *Adam* is usually translated "man" in this story (see 1.27 n.). Note that the curse is upon the ground, not man. **19:** Till you return to the ground: The mortal nature of man was implicit in the circumstances of his origin (2.7); because of man's disobedience, God now makes death an inevitable fate that haunts man throughout life. **21:** *Garments of skins*, a sign of God's protective care even in the time of judgement (4.15). **22:** *Like one of us*, see 3.5 n. *The tree of life* (2.9) does not figure in the temptation story, which explicitly speaks of only one tree in the center of the garden (3.3-6, 11-12, 17). **24:** *The Cherubim*, guardians of sacred areas (1 kg. 8.6-7), were represented as winged creatures like the sphynx of Egypt, half human and half lion (Ezek.41.18-19). *A flaming sword* (compare Jer.47.6) was placed near the cherubim to remind banished man of the impossibility of overstepping his creaturely bounds (compare Ezek.28.13-16).

28 *The Hebrew Bible*: Leviticus

[1]And the Lord said to Moses, [2]"Say to the people of Israel, I am the Lord your God. [3]You shall not do as they do in the land of Egypt, where you dwelt, and you shall not do as they do in the land of Canaan, to which I am bringing you. You shall not walk in their statutes. [4]You shall do my ordinances and keep my statutes and walk in them. I am the Lord your God. [5]You shall therefore keep my statutes and my ordinances, by doing which a man shall live: I am the Lord.

[6]"None of you shall approach any one near of kin to him to uncover nakedness. I am the Lord. [7]You shall not uncover the nakedness of your father, which is the nakedness of your mother; she is your mother, you shall not uncover her nakedness. [8]You shall not uncover the nakedness of your father's wife; it is your father's nakedness. [9]You shall not uncover the nakedness of your sister, the daughter of your father or the daughter of your mother, whether born at home or born abroad. [10]You shall not uncover the nakedness of your son's daughter or of your daughter's daughter, for their nakedness is your own nakedness. [11]You shall not uncover the nakedness of your father's wife's daughter, begotten by your father, since she is your sister. [12]You shall not uncover the nakedness of your father's sister; she is your father's near kinswoman. [13]You shall not uncover the nakedness of your mother's sister, for she is your mother's near kinswoman. [14]You shall not uncover the nakedness of your father's brother, that is, you shall not approach his wife; she is your aunt. [15]You shall not uncover the nakedness of your daughter-in-law; she is your son's wife, you shall not uncover her nakedness. [16]You shall not uncover the nakedness of your brother's wife; she is your brother's nakedness. [17]You shall not uncover the nakedness of a woman and of her daughter, and you shall not take her son's daughter or her daughter's daughter to uncover her nakedness; they are your near kinswomen; it is wickedness. [18]And you shall not take a woman as a rival wife to her sister, uncovering her nakedness while her sister is yet alive.

[19]"You shall not approach a woman to uncover her nakedness while she is in her menstrual uncleanness. [20]And you shall not lie carnally with your neighbor's wife, and defile yourself with her. [21]You shall not give any of your children to devote them by fire to Molech, and so profane the name of your God: I am the Lord. [22]You shall not lie with a male as with a woman; it is an abomination. [23]And you shall not lie with any beast and defile yourself with it, neither shall any woman give herself to a beast to lie with it: it is perversion.

[24]"Do not defile yourselves by any of these things, for by all these the nations I am casting out before you defiled themselves; [25]and the land became defiled, so that I punished its iniquity, and the land vomited out its inhabitants. [26]But you shall keep my statutes and my ordinances and do none of these abominations, either the native or the stranger who sojourns among you[27] (for all of these abominations the men of the land did, who were before you, so that the land became defiled); [28]lest the land vomit you out when you defile it, as it vomited out the nation that was before you. [29]For whoever shall do any of these abominations, the persons that do them shall be cut off from among their people. [30]So these abominable customs which were practised before you, and never to defile yourselves by them: I am the Lord your God."

End notes

f Gk: Heb lacks *your*

18.1-30: Forbidden sexual relations. 1-5: As a holy people, set apart for special relation to the Lord, Israel must not imitate the practices of other peoples (vv. 24-29; 11.44-45 n.)- **6-18:** An old list of twelve sexual prohibitions (compare the twelve curses in Dt. ch. 27). **16:** The levirate marriage was an exception to this rule (see Gen. 38.8 n.). **21:** On the pagan rite of child sacrifice to *Molech*, the Ammonite deity (1 Kg.11.7), see Dt.18.10 n. **26-28:** Although the laws of Leviticus are placed in the ancient setting of Mount Sinai, this passage clearly presupposes a time after the conquest of Canaan (vv. 25,27).

29 *Summa Theologica*

St. Thomas Aquinas

Question 153, 2 - SECOND ARTICLE
Whether No Venereal Act Can Be Without Sin

Let us move to the second article.
OBJECTION I It would seem that no venereal act can be sinless. For nothing but sin hinders virtue and every venereal act is a very definite hindrance to virtue. To quote Augustine, *"I do not consider anything more capable of sending a man's mind plummeting from its lofty heights than a woman's charm and the touch of their bodies."* Therefore it seems that no venereal act is sinless.
OBJECTION II Furthermore any excess that makes one forsake the good of reason is sinful. For we know from the *Ethics* of Aristotle that there are two ways of destroying virtue—excess and deficiency. Now in every sexual act there is an excess of pleasure which so absorbs the mind that it is impossible to think about anything else as Aristotle has noted. Or as St. Jerome says, while engaged in sex the prophets are out of the reach of every prophetic impulse. Therefore no venereal act can be sinless.
OBJECTION III The cause is more powerful than its effect. Now original sin is transmitted to children by concupiscence, without which no venereal act is possible as Augustine points out. Therefore no venereal act is sinless.

But on the other hand Augustine also states, *"This is an adequate rebuttal to heretics, if they can grasp it, that anything which does not violate nature, customs or law is not sinful."* And he says this in reference to the polygamy practices by the patriarchs. Therefore not every sexual act is a sin.

THE RESPONSE—In the realm of human actions, those which we properly call sins are the ones which violate the priorities of reason which arranges everything according to its proper end. Wherefore if the end is good and if what is done is in keeping with that end, and in harmony with it, then no sin is present. Now just as the preservation of a single human life is obviously good, how much greater is it to preserve the whole human race and just as the use of food is directed to the preservation of the life of the individual, so is the use of venereal acts ordered to the survival of the race. Hence Augustine says: *"What food is to man's well being, such is sexual intercourse to the well being of the race."* Wherefore just as the use of food can be without sin if it be taken in due manner and order as is required for the welfare of the body, so also the use of venereal acts can be without sin if they are performed in the proper manner and ordered to the preservation of the race.

REPLY—To the first objection it must be said that it is possible to hinder virtue in two ways, one by opposing it at a basic level which necessarily involves sin, and the other by hindering the highest expression of it by some action that is not a sin but merely a lesser good. And in this way sexual intercourse may send the mind plummeting not from virtue, but from virtue's highest expression. Hence Augustine says, *"Martha was doing good when she concerned herself with serving the holy men but Mary was doing a greater good by listening to the work of God. So too we praise the worth of Susanna's conjugal chastity but prefer the worth of the widow Anna and still more that of the Virgin Mary."*

2 To the second objection as stated above the balance of virtue is not attained by the yardstick of quantity but by the degree it is in harmony with the best dictates of reason. And therefore the exceeding pleasure experienced in the sex act, so long as it is in harmony with reason, does not destroy the balance of virtue. Furthermore virtue is not concerned with the amount of pleasure experienced by the external senses, as this depends on the disposition of the body; what matters is how much the interior appetite is affected by that pleasure. Nor must we conclude from the fact that reason cannot attend to spiritual things while experiencing venereal pleasure that the act itself is opposed to virtue any more that it would be safe to assume that any reasonable interruption of the activity of reason goes against virtue—otherwise it would be immoral to fall asleep. However this much is true, that sexual desire and pleasure are not subject to the sway and moderation of reason as a result of original sin; inasmuch as the reason for rebelling against God deserves to have its body rebel against it, as Augustine says."

3 To the third objection Augustine says, *"a child is barn shackled by original sin as a daughter of sin from the concupiscence of the flesh which in the newborn is not imputed to them as sins."* Hence it does not follow that the act in question is a sin but that it contains something penal derived from original sin.

THIRD ARTICLE

Whether the lust that surrounds the sexual act is sinful.

Let us proceed to the third article. It would seem that the lust which surrounds the sexual act is not sinful. For through the sexual act semen is discharged which is an excess by-product of food according to Aristotle. But there is no sin in the emission of other superfluous substances. Therefore, neither can be anything sinful concerning the sexual acts.

2 Furthermore everyone can lawfully use what is lawfully his own. But in the sexual act what is a man using except that which is his own except perhaps in the case of adultery or rape. Therefore there can be no sin in sexual activity and therefore lust would not be a sin.
3 Besides all sin has a vice which opposites it. But lust does not seem to have an opposite vice. Therefore lust is not a sin. On the other hand the cause is greater than its effect. But wine is forbidden by the Apostle Paul on account of lust, *"Do not allow yourselves to become drunk with wine wherein there is lust."* Therefore lust is prohibited.

Besides in Galatians it is enumerated among the works of the flesh.

THE RESPONSE—It must be said that the more a thing is necessary the more crucial it is that it be regulated by reason and consequently the greater the evil if that order of reason is violated. However as we have already stated sexual acts are very necessary for the common good which is the survival of the race. Therefore concerning them the greatest attention must be paid to the ordering of reason. Consequently if anything is done in connection with them against the order of reason it is a sin. And so without any doubt lust is sin.
1 To the first objection it must be said that according to Aristotle in the same book *semen is a surplus* which is necessary, he says it is a surplus insofar as it is a residue of the action of the nutrient power yet it is necessary regarding the generative power. The other waste-products of the body are not necessary and therefore it does not matter how they are disposed of provided the decencies of social life are observed. But it is different with the emission of semen which must be done in the way befitting the end for which it is needed.

To the second objection Paul speaks against lust, *"You were brought at a great price. Give glory and carry God in your bodies."* It follows from this that anyone who uses his body in an inordinate way because of lust offends God who is the Supreme Lord of our Bodies. Hence Augustine says that *"God who governs his servants to their own advantage commands them not to defile the temple of their bodies by illicit pleasures."*

To the third objection we reply that the opposite of lust is not frequently encountered since men are very prone to pleasure. Nevertheless, it is a vice that comes under the heading of insensibility and occurs in one who dislikes the sex relation so much that he doesn't even give his wife her conjugal rights.

Question 154
Whether We Can Aptly Classify Sins Of Lust Under Six Species

THE FIRST POINT—It would seem to be inappropriate to divide lust into six categories, simple fornication, adultery, incest, seduction, rape and unnatural sin. For diversity of material does not make for diverse species. But the aforementioned division is made on the basis of diversity of material, according to whether the act is committed by a woman who is married or unmarried or of some other circumstance. Therefore it would seem that the species of lust are not distinguishable on this basis.
2 Furthermore one species of vice should not be differentiated on the basis of something proper to another type of vice. But adultery differs from fornication only by the fact that

some takes to himself a woman who belongs to someone else and commits an injustice and so adultery should not be classified as a special or different type of lechery.

3 Again, just as it can happen that a man have relations with a woman who is bound to another man by marriage, so too he might have relationships with a woman who is bound to God by vow. Thus if adultery is considered a species of lust so too sacrilege should also be a species of lust.

4 Furthermore someone who is married not only sins if he goes to another woman but even if he uses his own wife inordinately. But this sin comes under the category of lechery and should therefore be counted as one of the types of lechery.

5 Furthermore Paul wrote, *"May God not humiliate me a second time when I come to you and bewail the fact that many of those who had sinned before still do not repent of the uncleanness, fornication and lewdness they have committed."* It seems therefore that uncleanness and lewdness must be considered species of lust just as fornication.

6 Furthermore one does not further separate a subject from the things it is divided into. Now we have mentioned the divisions of lechery, but Paul in Galatians says, *"The works of the flesh are manifest, they are fornication, uncleanness, impurity and lechery."* It would therefore seem inappropriate to make fornication a species of lechery.

On the other hand, this six-fold division was adopted in the *Decretum*.

REPLY—It should be mentioned, as we have already mentioned, that the sin of lechery consists in this—that a person engages in venereal pleasure not in accordance with right reason. This can happen in two ways: one in respect to the substance of the act in which the pleasure is sought. The other in which the act is proper but other conditions are not met. And because the circumstances of such an act do not determine the species of a moral act. But its species is only determined by its object which is the material of the act we must therefore assign separate species of these acts solely on the basis of material or objective.

Now it may conflict with right reason in two ways. One because it is repugnant to the purpose of the venereal act and thus, insofar as it impedes the generation of children, it is a vice against nature which is any venereal act from which generation cannot follow. However insofar as the necessary education and advancement of the offspring is impeded we have simple fornication between unmarried people.

The other way in which the sexual act is exercised against right reason arises from a comparison to other human beings and this can happen in two ways. First on the part of the woman herself with whom someone has relations because the respect due her is not given. Thus we have the sin of incest which consists in the abuse of a woman by one related by blood or affinity. The second way is on the part of the one who is in charge of her; if it be her husband, it is adultery; if her father, it is seduction (if force is not used) and rape if she is taken by force.

Notice however that these species of lechery differ more from considering the woman's role than the man's. Because in the sex act she is the passive one and for that reason the material element whereas the man is the agent. And as we mention, the different species are assigned on the basis of material differences.

1 To the first objection it must be said that the aforementioned diversity of material has a corresponding specific difference of objectives understood as the different ways it violates right reason.
2 To the second objection we reply that nothing prevents the several ugliness of various vices from occurring together in the same act and in this way adultery comes under the heading of both lust and injustice. Nor is injustice entirely incidental to injustice for lust becomes worse when following concupiscense it leads to injustices.
3 To the third we reply that because a woman who has vowed continence enters into a kind of Spiritual Marriage with God, sacrilege which is committed by violating such a woman is a type of spiritual adultery and similarly other kinds of sacrilege that involve lusting can be reduced to some type of lechery or other.
4 To the fourth objection we reply the sins of husbands with their wives are sin not because of improper material but because of attending circumstances and so, as was said, they do not constitute a specific kind of moral act.
5 To the fifth let me say that as the gloss says that uncleanncss stands for lust against nature, while lewdness pertains to corruption of the young which is a form of seduction. Or we can also say that lewdness pertains to those various acts surrounding the sex-act such as kissing, caressing and the like.
6 Finally to the sixth we respond that lust according to the dictionary includes any kind of excess.

ELEVENTH ARTICLE
Whether the Unnatural Vice Is a Species Of Lechery

Let us proceed to the eleventh article.
1 It would seem that vices against nature are not a type of lechery because in the aforementioned list there is no mention made of vice against nature. Therefore they are not a species of lechery.
2 Furthermore, lechery is opposed to the virtuous and is therefore a kind of wickedness. But unnatural vice is not a kind of human wickedness but a type of bestiality as Aristotle makes clear in his Ethics. Therefore a vice against nature is not human lust.
3 Besides lechery concerns activity that serves human generation. Unnatural sin is an act from which generation cannot follow. And therefore is not a kind of lechery.
On the other hand Second Corinthians numbers it among other types of lust. *"And they have done penance for the uncleanness and fornication and lasciviousness,"* where the gloss says *"lasciviousness is unnatural lust."*

I answer that as stated above wherever there occurs a special kind of deformity whereby the venereal act is rendered unbecoming, there is a determinate species of lust. This may occur in two ways: by being contrary to right reason and this is common to all types of lust. Secondly, because in addition it is contrary to the natural order of the venereal act as befits human species and this is why it is called the unnatural vice. It can happen in various ways. First by having orgasm outside of intercourse for the sake of venereal pleasure which pertains to the sin of uncleanness which some call effeminacy. Secondly by copulating with another species which is called bestiality. Thirdly by copulating with the wrong sex, male with male or female with female as St. Paul states (Rom. 1:27) and this is the vice of sodomy. Fourthly, by not observing the natural manner of copulation either as to not using the proper organ or as to other monstrous and bestial manners of copulation.

1 We reply to the first by saying, that there we were enumerating the species of lust that were not against human nature; wherefore the unnatural vices were omitted.

2 Bestiality goes beyond human wickedness for the latter is opposed to human virtue by a certain excess in regards to the same field and therefore can be reduced to the same genus.

3 The lustful man intends not human generation but venereal pleasure. It is possible without those acts from which human generation follow: and it is this which is sought in unnatural acts.

Question 154 - TWELFTH ARTICLE

Whether the Unnatural Vice Is the Greatest Sin Among the Species Of Lust?

We proceed thus to the Twelfth Article

OBJECTION I It seems that the unnatural vice is not the greatest sin among the species of lust. For the more a sin is contrary to charity the graver it is. Now adultery, seduction and rape which are injurious to our neighbor are seemingly more contrary to the love of our neighbor, than unnatural sins, by which no other person is injured. Therefore the unnatural sin is not the greatest among the species of lust.

OBJECTION II Further, sins committed against God would seem to be the most grievous. Now sacrilege is committed directly against God, since it is injurious to the Divine worship. Therefore sacrilege is a graver sin than the unnatural vice.

OBJECTION III Further, seemingly, a sin is all the more grievous according as we owe a greater love to the person against whom that sin is committed. Now the order of charity requires that a man love more those persons who are united to him—and such are those whom he defiles by incest,—than persons who are not connected with him, and whom in certain cases he defiles by the unnatural vice. Therefore, incest is a graver sin than the unnatural vice.

OBJECTION IV Further, if the unnatural vice is the most grievous, the more it is against nature the graver it would seem to be. Now the sin of uncleanness or effeminacy would seem to be most contrary to nature, since it would seem especially in accord with nature that agent and that which he acts upon should be distinct from one another. Hence it would follow that uncleanness is the gravest of unnatural vices. But this is not true. Therefore unnatural vices are not the most grievous among the sins of lust.

On the contrary, Augustine says (*De adult, conjug.*) that of all these vices (belonging, namely, to lust) that which is against nature is the worst.

I answer that, in every genus, worst of all is the corruption of the principle on which the rest depend. Now the principles of reason are those things that are according to nature because reason presupposes things as determined by nature, before disposing of other things as it is fitting. This may be observed both in speculative and in practical matters. Wherefore just as in speculative matters the most grievous and shameful error is that which is about things the knowledge of which is naturally bestowed on man, so in matters of action it is most grave and shameful to act against things as determined by nature. Therefore, since by the unnatural vices man transgresses that which has been determined by nature with regard to the use of venereal actions, it follows that in this matter this sin is gravest of sins. After which comes incest which, as has been said, offends against the natural respect we should have for those who are related to us.

Other kinds of lechery only pertain to rules arrived at by reason from principles of nature. For instance it is more repugnant to reason to have intercourse not only in a manner which not only harms the offspring but also injures the other person. Therefore, simple fornication which is committed without injuring the partner is the least among the sins of lechery. Furthermore, it is a greater sin to abuse married women than it is to abuse a woman who is under a guardian's care and therefore adultery is worse than seduction and if violence is also a factor the sin is worse, so it follows that raping a virgin is worse than seducing her and raping a wife is worse than adultery. All of these as we said above are made more serious offenses if sacrilege is also involved.

1 To the first objection we reply that men can fashion patterns of thought but God Himself arranged the natural orders. And so a sin against nature in which the natural order itself is violated is a sin against God who is the creator of that order. Augustine writes, *"Offenses against nature should be abhorred and punished always and in every case. Such as those committed by the people of Sodom which if every nation committed them they would be held just as guilty by the same divine law which never intended that men treat one another in such a fashion. For the fellowship which should exist between God and man would be destroyed when nature which is God's handiwork is made foul by the perversity of lust."*

2 To the second, it must be said that the sins against nature are against God himself, and in fact they are worse than sacrilege since the order of nature is more basic and stable than the laws which are derived from nature by reason.

3 To the third it will be said that the individual is more bound to the nature of species than he is to other individual members of that species and therefore the sin which attacks nature itself is more grievous.

4 Finally, sins of abuse are more serious than sins of omission. And so among the unnatural sins, masturbation holds the lowest because it omits the involvement of another person. The gravest, however is bestiality which does not even involve the same species. Thus on the text found in Genesis, *"He accused his brothers of the worst sin."* The commentary interprets this as meaning "they copulated with cattle." After bestiality, sodomy is the worst because the wrong sex is involved. After sodomy comes the sins of lechery which involve improper manners of intercourse and this is worse if it is not effected in the proper vessel than if the perversion of the sex act is done in some other way.

30 *Philosophy in the Bedroom*

Marquis De Sade

TO LIBERTINES

Voluptuaries of all ages, of every sex, it is to you only that I offer this work; nourish yourselves upon its principles: they favor your passions, and these passions, whereof coldly insipid moralists put you in fear, are naught but the means Nature employs to bring man to the ends she prescribes to him; harken only to these delicious promptings, for no voice save that of the passions can conduct you to happiness.

Lewd women, let the voluptuous Saint-Ange be your model; after her example, be heedless of all that contradicts pleasure's divine laws, by which all her life she was enchained.

You young maidens, too long constrained by a fanciful Virtue's absurd and dangerous bonds and by those of a disgusting religion, imitate the fiery Eugenie; be as quick as she to destroy, to spurn all those ridiculous precepts inculcated in you by imbecile parents.

And you, amiable debauchees, you who since youth have known no limits but those of your desires and who have been governed by your caprices alone, study the cynical Dolmancé, proceed like him and go as far as he if you too would travel the length of those flowered ways your lechery prepares for you; in Dolmancé's academy be at last convinced it is only by exploring and enlarging the sphere of his tastes and whims, it is only by sacrificing everything to the 'senses' pleasure that this individual, who never asked to be cast into this universe of woe, that this poor creature who goes under the name of Man, may be able to sow a smattering of roses atop the thorny path of life.

DIALOGUE THE FIFTH

Dolmancé, Le Chevalier, Augustin, Eugenié, Madame De Saint-Ange

36 year old man | 20 year old | 15 year old V | 26 year old woman

MADAME DE SAINT-ANGE, *presenting Augustin*—Let's on with it, friends, let's to our frolics; what would life be without its little amusments?...

EUGENIE, *blushing*—Heavens! I am so ashamed!

DOLMANCÉ—Rid yourself of that weak-hearted sentiment; all actions, and above all those of libertinage, being inspired in us by Nature, there is not one, of whatever kind, that warrants shame. Be smart there, Eugenie, act the whore with this young man; consider that every provocation sensed by a boy and originating from a girl is a natural offertory, and that your sex never serves Nature better than when it prostitutes itself to ours; that 'tis, in a word, to be fucked that you were born, and that she who refuses her obedience to this intention Nature has for her does not deserve to see the light longer....

DOLMANCÉ—Start from one fundamental point, Eugenie: in libertinage, nothing is frightful, because everything libertinage suggests is also a natural inspiration....

EUGENIE—Oh, 'tis natural?

DOLMANCÉ—Yes, natural, so I affirm it to be; Nature has not got two voices, you know, one of them condemning all day what the other commands, and it is very certain that it is nowhere but from her organ that those men who are infatuated with this mania receive the impressions that drive them to it. They who wish to denigrate the taste or proscribe its practice declare it is harmful to population; how dull-witted they are, these imbeciles who think of nothing but the multiplication of their kind, and who detect nothing but the crime in anything that conduces to a different end. Is it really so firmly established that Nature has so great a need for this overcrowding as they would like to have us believe? Is it very certain that one is guilty of an outrage whenever one abstains from this stupid propagation? To convince ourselves, let us for an instant scrutinize both her operations and her laws. Were it that Nature did naught but create, and never destroy, I might be able to believe, with those tedious sophists, that the sublimest of all actions would be incessantly to labor at production, and following that, I should grant, with them, that the refusal to reproduce would be, would perforce have to be, a crime; however, does not the most fleeting glance at natural operations reveal that destructions are just as necessary to her plan as are creations? That the one and the other of these functions are interconnected and enmeshed so intimately that for either to operate without the other would be impossible? That nothing would be born, nothing would be regenerated without destructions? Destruction, hence, like creation, is one of Nature's mandates....

But, the fools and the populators continue to object—and they are naught but one—this procreative sperm cannot have been placed in your loins for any purpose other than reproduction: to misuse it is an offense. I have just proven the contrary, since this misuse would not even be equivalent to destruction, and since destruction, far more serious than misuse, would not itself be criminal. Secondly, it is false that Nature intends this spermatic liquid to be employed only and entirely for reproduction; were this true, she would not permit its spillage under any circumstance save those appropriate to that end. But experience shows that the contrary may happen, since we lose it both when and where we wish. Secondly, she would forbid the occurrence of those losses save in coitus, losses which, however, do take place, both when we dream and when we summon remembrances; were Nature miserly about this so precious sap, 'twould never but be into the vessel of reproduction she would tolerate its flow; assuredly, she would not wish this voluptuousness, wherewith at such moments she crowns us, to be felt by us when we divert our tribute; for it would not be reasonable to suppose she could consent to give us pleasures at the very moment we heaped insults upon her. Let us go further; were women not born save to produce;—which most surely would be the case were this production so dear to Nature—, would it happen that, throughout the whole length of a woman's life, there are no more than seven years, all the arithmetic performed, during which she is in a state capable of conceiving and giving birth? What! Nature avidly seeks propagation, does she; and everything which does not tend to this end offends her, does it! And out of a hundred years of life the sex destined to produce cannot do so during more than seven years! Nature wishes for propagation only, and the semen she accords man to serve in these reproducings is lost, wasted, misused wherever and as often as it pleases man! He takes the same pleasures in this loss as in useful employment of his seed, and never the least inconvenience!...

Why, she would simply fail to notice it. Do you fancy races have not already become extinct? Buffon counts several of them perished, and Nature, struck dumb by a so precious loss, doesn't so much as murmur! The entire species might be wiped out and the air would not be the less pure for it, nor the Star less brilliant, nor the universe's march less exact. What idiocy it is to think that our kind is so useful to the world that he who might not labor to propagate it or he who might disturb this propagation would necessarily become a criminal! Let's bring this blindness to a stop and may the example of more reasonable peoples serve to persuade us of our errors. There is not one corner of the earth where the alleged crime of sodomy has not had shrines and votaries. The Greeks, who made of it, so to speak, a virtue, raised a statue unto Venus Callipygea; Rome sent to Athens for law, and returned with this divine taste... .

O my friends, can there be an extravagance to equal that of imagining that a man must be a monster deserving to lose his life because he has preferred enjoyment of the asshole to that of the cunt, because a young man with whom he finds two pleasures, those of being at once lover and mistress, has appeared to him preferable to a young girl, who promises him but half as much! He shall be a villain, a monster, for having wished to play the role of a sex not his own! Indeed! Why then has Nature created him susceptible of this pleasure?

MANNERS

After having made it clear that theism is in no wise suitable to a republican government, it seems to me necessary to prove that French manners are equally unsuitable to it. This article is the more crucial, for the laws to be promulgated will issue from manners, and will mirror them.

Frenchmen, you are too intelligent to fail to sense that new government will require new manners. That the citizens of a free State conduct themselves like a despotic king's slaves is unthinkable: the differences of their interests, of their duties, of their relations amongst one another essentially determine an entirely different manner of behaving in the world; a crowd of minor faults and of little social indelicacies, thought of as very fundamental indeed under the rule of kings whose expectations rose in keeping with the need they felt to impose curbs in order to appear respectable and unapproachable to their subjects, are due to become as nothing with us; other crimes with which we are acquainted under the names of regicide and sacrilege, in a system where kings and religion will be unknown, in the same way must be annihilated in a republican State. In according freedom of conscience and of the press, consider, citizens—for it is practically the same thing— whether freedom of action must not be granted too: excepting direct clashes with the underlying principles of government, there remain to you it is impossible to say how many fewer crimes to punish, because in fact there are very few criminal actions in a society whose foundations are liberty and equality. Matters well weighed and things closely inspected, only that is really criminal which rejects the law; for Nature, equally dictating vices and virtues to us, in reason of our constitution, yet more philosophically, in reason of the need Nature has of the one and the other, what she inspires in us would become a very reliable gauge by which to adjust exactly what is good and bad. But, the better to develop my thoughts upon so important a question, we will classify the different acts in man's life that until the present it has pleased us to call criminal, and we will next square them to the true obligations of a republican.

In every age, the duties of man have been considered under the following three categories:

1. Those his conscience and his credulity impose upon him, with what regards a supreme being;
2. Those he is obliged to fulfill toward his brethren;
3. Finally, those that relate only to himself....

I trust I have said enough to make plain that no laws ought to be decreed against religious crimes, for that which offends an illusion offends nothing, and it would be the height of inconsistency to punish those who outrage or who despise a creed or a cult whose priority to all others is established by no evidence whatsoever. No, that would necesarily be to exhibit a partiality and, consequently, to influence the scales of equality, that foremost law of your new government.

We move on to the second class of man's duties, those which bind him to his fellows; this is of all the classes the most extensive.

Excessively vague upon man's relations with his brothers, Christian morals propose bases so filled with sophistries that we are completely unable to accept them, since, if one is pleased to erect principles, one ought scrupulously to guard against founding them upon sophistries. This absurd morality tells us to love our neighbor as ourselves. Assuredly, nothing would be more sublime were it ever possible for what is false to be beautiful. The point is not at all to love one's brethren as oneself, since that is in defiance of all the laws of Nature, and since hers is the sole voice which must direct all the actions in our life; it is only a question of loving others as brothers, as friends given us by Nature, and with whom we should be able to live much better in a republican State, wherein the disappearance of distances must necessarily tighten the bonds.

May humanity, fraternity, benevolence prescribe but reciprocal obligations, and let us individually fulfill them with the simple degree of energy Nature has given us to this end; let us do so without blaming, and above all without punishing, those who, of chillier temper or more acrimonious humor, do not notice in these yet very touching social ties all the sweetness and gentleness others discover therein; for, it will be agreed, to seek to impose universal laws would be a palpable absurdity: such a proceeding would be as ridiculous as that of the general who would have all his soldiers dressed in a uniform of the same size; it is a terrible injustice to require that men of unlike character all be ruled by the same law: what is good for one is not at all good for another.

That we cannot devise as many laws as there are men must be admitted; but the laws can be lenient, and so few in number, that all men, of whatever character, can easily observe them. Furthermore, I would demand that this small number of laws be of such a sort as to be adaptable to all the various characters; they who formulate the code should follow the principle of applying more or less, according to the person in question. It has been pointed out that there are certain virtues whose practice is impossible for certain men, just as there are certain remedies which do not agree with certain constitutions. Now, would it not be to carry your injustice beyond all limits were you to send the law to strike the man incapable of bowing to the law? Would your iniquity be any less here than in a case where you sought to force the blind to distinguish amongst colors?

From these first principles there follows, one feels, the necessity to make flexible, mild laws and especially to get rid forever of the atrocity of capital punishment, because the law which attempts a man's life is impractical, unjust, inadmissible. Not, and it will be clarified in the sequel, that we lack an infinite number of cases where, without offense to Nature (and this I shall demonstrate), men have freely taken one another's lives, simply exercising a prerogative received from their common mother; but it is impossible for the law to obtain the same privileges, since the law, cold and impersonal, is a total stranger to the passions which are able to justify in man the cruel act of murder. Man receives his impressions from Nature, who is able to forgive him this act; the law, on the contrary, always opposed as it is to Nature and receiving nothing from her, cannot be authorized to permit itself the same extravagances: not having the same motives, the law cannot have the same rights. Those are wise and delicate distinctions which, escape many people, because very few of them reflect; but they will be grasped and retained by the instructed to whom I recommend them, and will, I hope, exert some influence upon the new code being readied for us....

The injuries we can work against our brothers may be reduced to four types: *calumny*; *theft*; the crimes which, caused by *impurity*, may in a disagreeable sense affect others; and *murder*...

Lay partiality aside, and answer me: is theft, whose effect is to distribute wealth more evenly, to be branded as a wrong in our day, under our government which aims at equality? Plainly, the answer is no: it furthers equality and, what is more, renders more difficult the conservation of property...

If, by your pledge, you perform an act of equity in protecting the property of the rich, do you not commit one of unfairness in requiring this pledge of the owner who owns nothing? What advantage does the latter derive from your pledge? And how can you expect him to swear to something exclusively beneficial to someone who, through his wealth, differs so greatly from him? Certainly, nothing is more unjust: an oath must have an equal effect upon all the individuals who pronounce it; that it bind him who has no interest in its maintenance is impossible, because it would no longer be a pact amongst free men; it would be the weapon of the strong against the weak, against whom the latter would have to be in incessant revolt.

Well, such, exactly, is the situation created by the pledge to respect property the Nation has just required all the citizens to subscribe to under oath. ...

Thus convinced, as you must be, of this barbarous inequality, do not proceed to worsen your injustice by punishing the man who has nothing for having dared to filch something from the man who has everything: your inequitable pledge gives him a greater right to it than ever. In driving him to perjury by forcing him to make a promise which, for him, is absurd, you justify all the crimes to which this perjury will impel him; it is not for you to punish something for which you have been the cause. I have no need to say more to make you sense the terrible cruelty of chastising thieves. ...

The transgressions we are considering in this second class of man's duties toward his fellows include actions for whose undertaking libertinage may be the cause; among those which are pointed to as particularly incompatible with approved behavior are *prostitution*, *incest*, *rape*, and *sodomy*. We surely must not for one moment doubt that all those known as moral crimes, that is to say, all acts of the sort to which those we have just cited belong, are of total inconsequence under a government whose sole duty consists in preserving, by whatever may be the means, the form essential to its continuance: there you have a republican government's unique morality. Well, the republic being permanently menaced from the outside by the despots surrounding it, the means to its preservation cannot be imagined as *moral means*, for the republic will preserve itself only by war, and nothing is less moral than war. I ask how one will be able to demonstrate that in a state rendered *immoral* by its obligations, it is essential that the individual be *moral*?...

Hence it would be no less absurd than dangerous to require that those who are to insure the perpetual *immoral* subversion of the established order themselves be *moral* beings: for the state of a moral man is one of tranquillity and peace, the state of an *immoral* man is one of perpetual unrest that pushes him to, and identifies him with, the necessary insurrection in which the republican must always keep the government of which he is a member....

If you would avoid that danger, permit a free flight and rein to those tyrannical desires which, despite himself, torment man ceaselessly: content with having been able to exercise his small dominion in the middle of the harem of sultanas and youths whose submission your good offices and his money procure for him, he will go away appeased and with nothing but fond feelings for a government which so obligingly affords him every means of satisfying his concupiscence; proceed, on the other hand, after a different fashion, between the citizen and those objects of public lust raise the ridiculous obstacles in olden times invented by ministerial tyranny and by the lubricity of our Sardanapaluses[1]—, do that, and the citizen, soon embittered against your regime, soon jealous of the despotism he sees you exercise all by yourself, will shake off the yoke you lay upon him, and, weary of your manner of ruling, will, as he has just done, substitute another for it....

I am going to try to convince you that the prostitution of women who bear the name of honest is no more dangerous than the prostitution of men, and that not only must we associate women with the lecheries practiced in the houses I have set up, but we must even build some for them, where their whims and the requirements of their temper, ardent like ours but in a quite different; may too find satisfaction with every sex...

Never may an act of possession be exercised upon a free being; the exclusive possession of a woman is no less unjust than the possession of slaves; men are born free, all have equal rights: never should we lose sight of those principles; according to which never may there be granted to one sex the legitimate right to lay monopolizing hands upon the other, and never may one of these sexes, or classes, arbitrarily possess the other. Similarly, a woman existing in the purity of Nature's laws cannot allege, as justification for refusing herself to someone who desires her, the love she bears another, because such a response is based upon exclusion, and no man may be excluded from the having of a woman as of the moment it is clear she definitely belongs to all men. The act of possession can only be exercised upon chattel or an animal, never upon an individual who resembles us, and all the ties which can bind a woman to a man are quite as unjust as illusory.

If then it becomes incontestable that we have received from Nature the right indiscriminately to express our wishes to all women, it likewise becomes incontestable that we have the right to compel their submission, not exclusively, for I should then be contradicting myself but temporarily.[2] It cannot be denied that we have the right to decree laws that compel woman to yield to the flames of him who would have her; violence itself being one of that right's effects, we can employ it lawfully. Indeed! Has Nature not proven that we have that right, by bestowing upon us the strength needed to bend women to our will?...

If we admit, as we have just done, that all Women ought to be subjugated to our desires, we may certainly allow then ample satisfaction of theirs. Our laws must be favorable to their fiery temperament. It is absurd to locate both their honor and their virtue in the antinatural strength they employ to resist the penchants with which they have been far more profusely endowed than we; this injustice of manners is rendered more flagrant still since we contrive at once to weaken them by seduction, and then to punish them for yielding to all the efforts we have made to provoke their fall. All the absurdity of our manners, it seems to me, is graven in this shocking paradox, and this brief outline alone ought to awaken us to the urgency of exchanging them for manners more pure...

Is incest more dangerous? Hardly. It loosens family ties and the citizen has that much more love to lavish on his country; the primary laws of Nature dictate it to us, our feelings vouch for the fact; and nothing is so enjoyable as an object we have coveted over the years. ... I would venture, in a word, that incest ought to be every government's law—every government whose basis is fraternity. How is it that reasonable men were able to carry absurdity to the point of believing that the enjoyment of ones mother, sister, or daughter could ever be criminal? Is it not, I ask, an abominable view wherein it is made to appear a crime for a man to place higher value upon the enjoyment of an object to which natural feeling draws him close? One might just as well say that we are forbidden to love too much the individuals Nature enjoins us to love best, and that the more she gives us a hunger for some object, the more she orders us away from it.... We shall turn our attention to rape, which at first glance seems to be of all libertinage's excesses, the one which is most dearly established as being wrong, by reason of the outrage it appears to cause. It is certain, however, that rape, an act so very rare and so very difficult to prove, wrongs ones neighbor less than theft, since the latter is destructive to property, the former merely damaging to it. Beyond that, what objections have you to the ravisher? What will you say, when he replies to you that, as a matter of fact, the injury he has committed is trifling indeed, since he has done no more than place a little sooner the object he has abused in the very state in which she would soon have been put by marriage and love.

But sodomy, that alleged crime which will draw the fire of heaven upon cities addicted to it...

What single crime can exist here? For no one will wish to maintain that all the parts of the body do not resemble each other, that there are some which are pure, and others defiled; but, as it is unthinkable such nonsense be advanced seriously, the only possible crime would consist in the waste of semen...

End notes

1 It is well known that the infamous and criminal Sartinc devised, in the interests of the king's lewdness, the plan of having Dubarry read to Louis XV, thrice each week, the private details, enriched by Sartine, or all that transpired in the evil corners of Paris. This department of the French Nero's libertinage cost the State three millions.

2 Let it not be said that I contradict myself here, arid that after having established, at some point further above, that we have no right to bind a woman to ourselves, I destroy those principles when I declare now we have the right to constrain her; I repeat,- it is a question of enjoyment only, not of property: I have no right of possession upon that fountain I find by the road, but I have certain rights to its use; I have the right to avail myself of the limpid water it offers my thirst; similarly, I have no real right of possession over such-and-such a woman, but I have incontestable rights to the enjoyment of her, I have the right to force from her this enjoyment, if she refuses me it for whatever the cause may be.

31 "Sexual Perversion"

Thomas Nagel

There is something to be learned about sex from the fact that we possess a concept of sexual perversion. I wish to examine the idea, defending it against the charge of unintelligibility and trying to say exactly what about human sexuality qualifies it to admit of perversions. Let me begin with some general conditions that the concept must meet if it is to be viable at all. These can be accepted without assuming any particular analysis.

First, if there are any sexual perversions, they will have to be sexual desires or practices that are in some sense unnatural, though the explanation of this natural/unnatural distinction is of course the main problem. Second, certain practices will be perversions if anything is, such as shoe fetishism, bestiality, and sadism; other practices, such as unadorned sexual intercourse, will not be; about still others there is controversy. Third, if there are perversions, they will be unnatural sexual *inclinations* rather than just unnatural practices adopted not from inclination but for other reasons. Thus contraception, even if it is thought to be a deliberate perversion of the sexual and reproductive functions, cannot be significantly described as a sexual perversion. A sexual *perversion* must reveal itself in conduct that expresses an unnatural *sexual* preference. And although there might be a form of fetishism focused on the employment of contraceptive devices, that is not the usual explanation for their use.

The connection between sex and reproduction has no bearing on sexual perversion. The latter is a concept of psychological not physiological, interest, and it is a concept that we do not apply to the lower animals, let alone to plants, all of which have reproductive functions that can go astray in various ways. (Think of seedless oranges.) Insofar as we are prepared to regard higher animals as perverted, it is because of their psychological, not their anatomical, similarity to humans. Furthermore, we do not regard as a perversion every deviation from the reproductive function of sex in humans: sterility, miscarriage, contraception, abortion.

Nor can the concept of sexual perversion be defined in terms of social disapprobation or custom. Consider all the societies that have frowned upon adultery and fornication. These have not been regarded as unnatural practices, but have been thought objectionable in other ways. What is regarded as unnatural admittedly varies from culture to culture, but the classification is not a pure expression of disapproval or distaste. In fact it is often regarded as a *ground* for disapproval, and that suggests that the classification has independent content.

I shall offer a psychological account of sexual perversion that depends on a theory of sexual desire and human sexual interactions. To approach this solution I shall first consider a contrary position that would justify skepticism about the existence of any sexual perversions at all, and perhaps even about the significance of the term. The skeptical argument runs as follows:

> 'Sexual desire is simply one of the appetites, like hunger and thirst. As such it may have various objects, some more common than others perhaps, but none in any sense "natural." An appetite is identified as sexual by means of the organs and erogenous zones in which its satisfaction can be to some extent localized, and the special sensory pleasures which form the core of that satisfaction. This enables us to recognize widely divergent goals, activities, and desires as sexual, since it is conceivable in principle that anything should produce sexual pleasure and that a nondeliberate, sexually charged desire for it should arise (as a result of conditioning, if nothing else). We may fail to empathize with some of these desires, and some of them, like sadism, may be objectionable on extraneous grounds, but once we have observed that they meet the criteria for being sexual, there is nothing more to be said on *that* score. Either they are sexual or they are not: sexuality does not admit of imperfection, or perversion, or any other such qualification - it is not that sort of affection.'

This is probably the received radical position. It suggests that the cost of defending a psychological account may be to deny that sexual desire is an appetite. But insofar as that line of defense is plausible, it should make us suspicious of the simple picture of appetites on which the skepticism depends. Perhaps the standard appetites, like hunger, cannot be classed as pure appetites in that sense either, at least in their human versions.

Can we imagine anything that would qualify as a gastronomical perversion? Hunger and eating, like sex, serve a biological function and also play a significant role in our inner lives. Note that there is little temptation to describe as perverted an appetite for substances that are not nourishing: we should probably not consider someone's appetites *perverted* if he liked to eat paper, sand, wood, or cotton. Those are merely rather odd and very unhealthy tastes: they lack the psychological complexity that we expect of perversions. (Coprophilia, being already a sexual perversion, may be disregarded.) If on the other hand someone liked to eat cookbooks, or magazines with pictures of food in them, and preferred these to ordinary food — or if when hungry he sought satisfaction by fondling a napkin or ashtray from his favorite restaurant - then the concept of perversion might seem appropriate (it would be natural to call it gastronomical fetishism). It would be natural to describe as gastronomically perverted someone who could eat only by having food forced down his throat through a funnel, or only if the meal were a living animal. What helps is the peculiarity of the desire itself, rather than the inappropriateness of its object to the biological function that the desire serves. Even an appetite can have perversions if in addition to its biological function it has a significant psychological structure.

In the case of hunger, psychological complexity is provided by the activities that give it expression. Hunger is not merely a disturbing sensation that can be quelled by eating; it is an attitude toward edible portions of the external world, a desire to treat them in rather special ways. The method of ingestion: chewing, savoring, swallowing, appreciating the texture and smell, all are important components of the relation, as is the passivity and controllability of the food (the only animals we eat live are helpless mollusks). Our relation to food depends also on our size: we do not live upon it or burrow into it like aphids or worms. Some of these features are more central than others, but an adequate phenomenology of eating would have to treat it as a relation to the external world and a way of appropriating bits of that world, with characteristic affection. Displacements or serious restrictions of the desire to eat could then be described as perversions, if they undermined that direct relation between man and food which is the natural expression of hunger. This explains why it is easy to imagine gastronomical fetishism, voyeurism, exhibitionism, or even gastronomical sadism and Masochism. Some of these perversions are fairly common.

If we can imagine perversions of an appetite like hunger, it should be possible to make sense of the concept of sexual perversion. I do not wish to imply that sexual desire is an appetite - only that being an appetite is no bar to admitting of perversions. Like hunger, sexual desire has as its characteristic object a certain relation with something in the external world; only in this case it is usually a person rather than an omelet, and the relation is considerably more complicated. This added complication allows scope for correspondingly complicated perversions.

The fact that sexual desire is a feeling about other persons may encourage a pious view of its psychological content - that it is properly the expression of some other attitude, like love, and that when it occurs by itself it is incomplete or subhuman. (The extreme Platonic version of such a view is that sexual practices are all vain attempts to express something they cannot in principle achieve: this makes them all perversions, in a sense.) But sexual desire is complicated enough without having to be linked to anything else as a condition for phenomenological analysis. Sex may serve various functions — economic, social, altruistic — but it also has its own content as a relation between persons.

The object of sexual attraction is a particular individual, who transcends the properties that make him attractive. When different persons are attracted to a single person for different reasons — eyes, hair, figure, laugh, intelligence — we nevertheless feel that the object of their desire is the same. There is even an inclination to feel that this is so if the lovers have different sexual aims, if they include both men and women, for example. Different specific attractive characteristics seem to provide enabling conditions for the operation of a single basic feeling, and the different aims all provide expressions of it. We approach the sexual attitude toward the person through the features that we find attractive, but these features are not the objects of that attitude.

This is very different from the case of an omelet. Various people may desire it for different reasons, one for its fluffiness, another for its mushrooms, another for its unique combination of aroma and visual aspect; yet we do not enshrine the transcendental omelet as the true common object of their affections. Instead we might say that several desires have accidentally converged on the same object: any omelet with the crucial characteristics would do as well. It is not similarly true that any person with the same flesh distribution and way of smoking can be substituted as object for a particular sexual desire that has been elicited by those characteristics. It may be that they recur, but it will be a new sexual attraction with a new particular object, not merely a transfer of the old desire to someone else. (This is true even in cases where the new object is unconsciously identified with a former one.)

The importance of this point will emerge when we see how complex a psychological interchange constitutes the natural development of sexual attraction. This would be incomprehensible if its object were not a particular person, but rather a person of a certain *kind*. Attraction is only the beginning, and fulfillment does not consist merely of behaviour and contact expressing this attraction, but involves much more.

The best discussion of these matters that I have seen appears in part III of Sartre's *Being and Nothingness*.[1] Sartre's treatment of sexual desire and of love, hate, sadism, masochism, and further attitudes toward others, depends on a general theory of consciousness and the body which we can neither expound nor assume here. He does not discuss perversion, and this is partly because he regards sexual desire as one form of the perpetual attempt of an embodied consciousness to come to terms with the existence of others, an attempt that is as doomed to fail in this form as it is in any of the others, which include sadism and masochism (if not certain of the more impersonal deviations) as well as several nonsexual attitudes. According to Sartre, all attempts to incorporate the other into my world as another subject, i.e. to apprehend him at once as an object for me and as a subject for whom I am an object, are unstable and doomed to collapse into one or other of the two aspects. Either I reduce him entirely to an object, in which case his subjectivity escapes the possession or appropriation I can extend to that object; or I become merely an object for him, in which case I am no longer in a position to appropriate his subjectivity. Moreover, neither of these aspects is stable; each is continually in danger of giving way to the other. This has the consequence that there can be no such thing as a *successful* sexual relation, since the deep aim of sexual desire cannot in principle be accomplished. It seems likely, therefore, that the view will not permit a basic distinction between successful or complete and unsuccessful or incomplete sex, and therefore cannot admit the concept of perversion.

I do not adopt this aspect of the theory, nor many of its metaphysical underpinnings. What interests me is Sartre's picture of the attempt. He says that the type of possession that is the object of sexual desire is carried out by 'a double reciprocal incarnation' and that this is accomplished, typically in the form of a caress, in the following way: 'I make myself flesh in order to impel the Other to realize *for herself* and *for me* her own flesh, and my caresses cause my flesh to be born for me in so far as it is for the *Other flesh causing her to be born as flesh*' (*Being and Nothingness*, p.391; Satre's italics). The incarnation in question is described variously as a clogging or troubling of consciousness, which is inundated by the flesh in which it is embodied.

The view I am going to suggest, I hope in less obscure language, is related to this one, but it differs from Sartre's in allowing sexuality to achieve its goal on occasion and thus in providing the concept of perversion with a foothold.

Sexual desire involves a kind of perception, but not merely a single perception of its object, for in the paradigm case of mutual desire there is a complex system of superimposed mutual perceptions - not only perceptions of the sexual object, but perceptions of oneself. Moreover, sexual awareness of another involves considerable self-awareness to begin with — more than is involved in ordinary sensory perception. The experience is felt as an assault on oneself by the view (or touch, or whatever) of the sexual object.

Let us consider a case in which the elements can be separated. For clarity we will restrict ourselves initially to the somewhat artificial case of desire at a distance. Suppose a man and a woman, whom we may call Romeo and Juliet, are at opposite ends of a cocktail lounge, with many mirrors on the walls which permit unobserved observation, and even mutual unobserved observation. Each of them is sipping a martini and studying other people in the mirrors. At some point Romeo notices Juliet. He is moved, somehow, by the softness of her hair and the diffidence with which she sips her martini, and this arouses him sexually. Let us say that *X senses Y* whenever *X* regards *Y* with sexual desire. (*Y* need not be a person, and *X's* apprehension of *Y* can be visual, tactile, olfactory, etc., or purely imaginary; in the present example we shall concentrate on vision.) So Romeo senses Juliet, rather than merely noticing her. At this stage he is aroused by an unaroused object, so he is more in the sexual grip of his body than she of hers.

Let us suppose, however, that Juliet now senses Romeo in another mirror on the opposite wall, though neither of them yet knows that he is seen by the other (the mirror angles provide three-quarter views). Romeo then begins to notice in Juliet the subtle signs of sexual arousal, heavy-lidded stare, dilating pupils, faint flush, etc. This of course intensifies her bodily presence, and he not only notices but senses this as well. His arousal is nevertheless still solitary. But now, cleverly calculating the line of her stare without actually looking her in the eyes, he realizes that it is directed at him through the mirror on the opposite wall. That is, he notices, and moreover senses, Juliet sensing him. This is definitely a new development, for it gives him a sense of embodiment not only through his own reactions but through the eyes and reactions of another. Moreover, it is separable from the initial sensing of Juliet; for sexual arousal might begin with a person's sensing that he is sensed and being assailed by the perception of the other person's desire rather than merely by the perception of the person.

But there is a further step. Let us suppose that Juliet, who is a little slower than Romeo, now senses that he senses her. This puts Romeo in a position to notice, and be aroused by, her arousal at being sensed by him. He senses that she senses that he senses her. This is still another level of arousal, for he becomes conscious of his sexuality through his awareness of its effect on her and of her awareness that this effect is due to him. Once she takes the same step and senses that he senses her sensing him, it becomes difficult to state, let alone imagine, further iterations, though they may be logically distinct. If both are alone, they will presumably turn to look at each other directly, and the proceedings will continue on another plane. Physical contact and intercourse are natural extensions of this complicated visual exchange, and mutual touch can involve all the complexities of awareness present in the visual case, but with a far greater range of subtlety and acuteness.

Ordinarily, of course, things happen in a less orderly fashion — sometimes in a great rush — but I believe that some version of this overlapping system of distinct sexual perceptions and interactions is the basic framework of any full-fledged sexual relation and that relations involving only part of the complex are significantly incomplete. The account is only schematic, as it must be to achieve generality. Every real sexual act will be psychologically far more specific and detailed, in ways that depend not only on the physical techniques employed and on anatomical details, but also on countless features of the participants' conceptions of themselves and of each other, which become embodied in the act. (It is familiar enough fact, for example, that people often take their social roles and the social roles of their partners to bed with them.)

The general schema is important, however, and the proliferation of levels of mutual awareness it involves is an example of a type of complexity that typifies human interactions. Consider aggression, for example. If I am angry with someone, I want to make him feel it, either to produce self-reproach by getting him to see himself through the eyes of my anger, and to dislike what he sees - or else to produce reciprocal anger or fear, by getting him to perceive my anger as a threat or attack. What I want will depend on the details of my anger, but in either case it will involve a desire that the object of that anger be aroused. This accomplishment constitutes the fulfillment of my emotion, through domination of the object's feelings.

Another example of such reflexive mutual recognition is to be found in the phenomenon of meaning, which appears to involve an intention to produce a belief or other effect in another by bringing about his recognition of one's intention to produce that effect. (That result is due to H. P. Grice,[2] whose position I shall not attempt to reproduce in detail.) Sex has a related structure: it involves a desire that one's partner be aroused by the recognition of one's desire that he or she be aroused.

It is not easy to define the basic types of awareness and arousal of which these complexes are composed, and that remains a lacuna in this discussion. In a sense, the object of awareness is the same in one's own case as it is in one's sexual awareness of another, although the two awarenesses will not be the same, the difference being as great as that between feeling angry and experiencing the anger of another. All stages of sexual perception are varieties of identification of a person with his body. What is perceived is one's own or another's *subjection* to or *immersion* in his body, a phenomenon which has been recognized with loathing by St Paul and St Augustine, both of whom regarded 'the law of sin which is in my members' as a grave threat to the dominion of the holy will.[3] In sexual desire and its expression the blending of involuntary response with deliberate control is extremely important. For Augustine, the revolution launched against him by his body is symbolized by erection and the other involuntary physical components of arousal. Sartre too stresses the fact that the penis is not a prehensile organ. But mere involuntariness characterizes other bodily processes as well. In sexual desire the involuntary responses are combined with submission to spontaneous impulses: not only one's pulse and secretions but one's actions are taken over by the body; ideally, deliberate control is needed only to guide the expression of those impulses. This is to some extent also true of an appetite like hunger, but the takeover there is more localized, less pervasive, less extreme. One's whole body does not become saturated with hunger as it can with desire. But the most characteristic feature of a specifically sexual immersion in the body is its ability to fit into the complex of mutual perceptions that we have described.

Hunger leads to spontaneous interactions with food; sexual desire leads to spontaneous interactions with other persons, whose bodies are asserting their sovereignty in the same way, producing involuntary reactions and spontaneous impulses in them. These reactions are perceived, and the perception of them is perceived, and that perception is in turn perceived; at each step the domination of the person by his body is reinforced, and the sexual partner becomes more possessible by physical contact, penetration, and envelopment.

Desire is therefore not merely the perception of a pre-existing embodiment of the other, but ideally a contribution to his further embodiment which in turn enhances the original subject's sense of himself. This explains why it is important that the partner be aroused, and not merely aroused, but aroused by the awareness of one's desire. It also explains the sense in which desire has unity and possession as its object: physical possession must eventuate in creation of the sexual object in the image of one's desire, and not merely in the object's recognition of that desire, or in his or her own private arousal.

Even if this is a correct model of the adult sexual capacity, it is not plausible to describe as perverted every deviation from it. For example, if the partners in heterosexual intercourse indulge in private heterosexual fantasies, thus avoiding recognition of the real partner, that would, on this model, constitute a defective sexual relation. It is not, however, generally regarded as a perversion. Such examples suggest that a simple dichotomy between perverted and unperverted sex is too crude to organize the phenomena adequately.

Still, various familiar deviations constitute truncated or incomplete versions of the complete configuration, and may be regarded as perversions of the central impulse. If sexual desire is prevented from taking its full interpersonal form, it is likely to find a different one. The concept of perversion implies that a normal sexual development has been turned aside by distorting influences. I have little to say about this causal condition. But if perversions are in some sense unnatural, they must result from interference with the development of a capacity that is there potentially.

It is difficult to apply this condition, because environmental factors play a role in determining the precise form of anyone's sexual impulse. Early experiences in particular seem to determine the choice of a sexual object. To describe some causal influences as distorting and others as merely formative is to imply that certain general aspects of human sexuality realize a definite potential whereas many of the details in which people differ realize an indeterminate potential, so that they cannot be called more or less natural. What is included in the definite potential is therefore very important, although the distinction between definite and indeterminate potential is obscure. Obviously a creature incapable of developing the levels of interpersonal sexual awareness I have described could not be deviant in virtue of the failure to do so. (Though even a chicken might be called perverted in an extended sense if it had been conditioned to develop a fetishistic attachment to a telephone.) But if humans will tend to develop some version of reciprocal interpersonal sexual awareness unless prevented, then cases of blockage can be called unnatural or perverted.

Some familiar deviations can be described in this way. Narcissistic practices and intercourse with animals, infants, and inanimate objects seem to be stuck at some primitive version of the first stage of sexual feeling. If the object is not alive, the experience is reduced entirely to an awareness of one's own sexual embodiment. Small children and animals permit awareness of the embodiment of the other, but present obstacles to reciprocity, to the recognition by the sexual object of the subject's desire as the source of his (the object's) sexual self-awareness. Voyeurism and exhibitionism are also incomplete relations. The exhibitionist wishes to display his desire without needing to be desired in return; he may even fear the sexual attentions of others. A voyeur, on the other hand, need not require any recognition by his object at all: certainly not a recognition of the voyeur's arousal.

On the other hand, if we apply our model to the various forms that may be taken by two-party heterosexual intercourse, none of them seem clearly to qualify as perversions. Hardly anyone can be found these days to inveigh against oral-genital contact, and the merits of buggery are urged by such respectable figures as D. H. Lawrence and Norman Mailer. In general, it would appear that any bodily contract between a man and a woman that gives them sexual pleasure is a possible vehicle for the system of multi-level interpersonal awareness that I have claimed is the basic psychological content of sexual interaction. Thus a liberal platitude about sex is upheld.

The really difficult cases are sadism, masochism, and homosexuality. The first two are widely regarded as perversions and the last is controversial. In all three cases the issue depends partly on causal factors: do these dispositions result only when normal development has been prevented? Even the form in which this question has been posed is circular, because of the word 'normal'. We appear to need an independent criterion for a distorting influence, and we do not have one.

It may be possible to class sadism and masochism as perversions because they fall short of interpersonal reciprocity. Sadism concentrates on the evocation of passive self-awareness in others, but the sadist's engagement is itself active and requires a retention of deliberate control which may impede awareness of himself as a bodily subject of passion in the required sense. De Sade claimed that the object of sexual desire was to evoke involuntary responses from one's partner, especially audible ones. The infliction of pain is no doubt the most efficient way to accomplish this, but it requires a certain abrogation of one's own exposed spontaneity. A masochist on the other hand imposes the same disability on his partner as the sadist imposes on himself. The masochist cannot find a satisfactory embodiment as the object of another's sexual desire, but only as the object of his control. He is passive not in relation to his partner's passion but in relation to his nonpassive agency. In addition, the subjection to one's body characteristic of pain and physical restraint is of a very different kind from that of sexual excitement: pain causes people to contract rather than dissolve. These descriptions may not be generally accurate. But to the extent that they are, sadism and masochism would be disorders of the second stage of awareness — the awareness of oneself as an object of desire.

Homosexuality cannot similarly be classed as a perversion on phenomenological grounds. Nothing rules out the full range of interpersonal perceptions between persons of the same sex. The issue then depends on whether homosexuality is produced by distorting influences that block or displace a natural tendency to heterosexual development. And the influences must be more distorting than those which lead to a taste for large breasts or fair hair or dark eyes. These also are contingencies of sexual preference in which people differ, without being perverted.

The question is whether heterosexuality is the natural expression of male and female sexual dispositions that have not been distorted. It is an unclear question, and I do not know how to approach it. There is much support for an aggressive-passive distinction between male and female sexuality. In our culture the male's arousal tends to initiate the perceptual exchange, he usually makes the sexual approach, largely controls the course of the act, and of course penetrates whereas the woman receives. When two men or two women engage in intercourse they cannot both adhere to these sexual roles. But a good deal of deviation from them occurs in heterosexual intercourse. Women can be sexually aggressive and men passive, and temporary reversals of role are not uncommon in heterosexual exchanges of reasonable length. For these reasons it seems to be doubtful that homosexuality must be a perversion, though like heterosexuality it has perverted forms.

Let me close with some remarks about the relation of perversion to good, bad, and morality. The concept of perversion can hardly fail to be evaluative in some sense, for it appears to involve the notion of an ideal or at least adequate sexuality which the perverions in some way fail to achieve. So, if the concept is viable, the judgment that a person or practice or desire is perverted will constitute a sexual evaluation, implying that better sex, or a better specimen of sex, is possible. This in itself is a very weak claim, since the evaluation might be in a dimension that is of little interest to us. (Though, if my account is correct, that will not be true.)

Whether it is a moral evaluation, however, is another question entirely — one whose answer would require more understanding of both morality and perversion than can be deployed here. Moral evaluation of acts and of persons is a rather special and very complicated matter, and by no means all our evaluations of persons and their activities are moral evaluations. We make judgments about people's beauty or health or intelligence which are evaluative without being moral. Assessments of their sexuality may be similar in that respect.

Furthermore, moral issues aside, it is not clear that unperverted sex is necessarily *preferable* to the perversions. It may be that sex which receives the highest marks for perfection *as sex* is less enjoyable than certain perversions; and if enjoyment is considered very important, that might outweigh considerations in determining rational preference.

That raises the question of the relation between the evaluative content of judgments of perversion and the rather common *general* distinction between good and bad sex. The latter distinction is usually confined to sexual acts, and it would seem, within limits, to cut across the other: even someone who believed, for example, that homosexuality was a perversion could admit a distinction between better and worse homosexual sex, and might even allow that good homosexual sex could be better *sex* than not very good unperverted sex. If this is correct, it supports the position that, if judgments of perversion are viable at all, they represent only one aspect of the possible evaluation of sex, even *qua sex*. Moreover it is not the only important aspect: sexual deficiencies that evidently do not constitute perversions can be the object of great concern.

Finally, even if perverted sex is to that extent not so good as it might be, bad sex is generally better than none at all. This should not be controversial: it seems to hold for other important matters, like food, music, literature, and society. In the end, one must choose from among the available alternatives, whether their availability depends on the environment or on one's own constitution. And the alternatives have to be fairly grim before it becomes rational to opt for nothing.

End notes

1 *L'Etre et le Neant* (Paris: Gallimand, 1943), translated by Hazel E. Barnes (New York: Philosophical Library, 1956).

2 'Meaning'. *Philosophical Review*, lxvi, no. 3 (July, 1957). 377-88.

3 Sec *Romans*, VII, 23; and the *Confessions*, bk vIII, pt V.

32 "The Human Sexual Response Cycle"

Sinclaire Intimacy Institute

SEXUAL RESPONSE

Sexual response refers to the set of physiological and emotional changes that lead to and follow orgasm. Different researchers have constructed various models.

Usually, these models include three, four, or five distinct phases, with the exact components of each phase differing across models.

Helen Singer Kaplan proposed the Triphasic Concept of human sexual response involving three stages: desire, excitement, and orgasm. In his book "Human Sexual Response", Lief described five sexual response phases: desire, arousal, vasocongestion, orgasm, and satisfaction.

William Masters and Virginia Johnson, prominent sex researchers and therapists, suggested that there are four identifiable phases in the sex response cycle: excitement, plateau, orgasm, and resolution.

Using various instruments designed to monitor changes in heart rate and muscle tension, Masters and Johnson were able specify the bodily changes that characterize each of these phases.

THE FIRST PHASE OF SEXUAL RESPONSE

Excitement can last for just a few minutes or extend for several hours. Characteristics of this phase include: an increasing level of muscle tension, a quickened heart rate, flushed skin (or some blotches of redness may occur on the chest and back), hardened or erect nipples, and the onset of vasocongestion, resulting in swelling of the woman's clitoris and labia minora and erection of the man's penis.

Other changes also occur. In the woman, the vaginal walls begin to produce a lubricating liquid, her uterus elevates and grows in size, and her breasts become larger. At the same time, the woman's vagina swells and the muscle that surrounds the vaginal opening, called the pubococygeal muscle, grows tighter.

These changes prepare the woman's body for orgasm and were called the "orgasmic platform" by Masters and Johnson. Additional changes in men include elevation and swelling of the testicles, tightening of the scrotal sac, and secretion of a lubricating liquid by the Cowper's glands.

THE SECOND PHASE OF SEXUAL RESPONSE

Also known as the plateau, this phase is characterized primarily by the intensification of all of the changes begun during the excitement phase. During this period, the woman's clitoris may become so sensitive that it is painful to the touch. The plateau phase extends to the brink of orgasm, which initiates the reversal of all of the changes begun during the excitement phase.

During the second phase, called the expulsion phase, the urinary bladder closes to block the possibility of urine mixing with the semen. At this point, muscles at the base of the penis begin a steady rhythmic contraction that finally expels the semen from the urethral opening at the head of the penis.

THE THIRD PHASE OF SEXUAL RESPONSE

The peak of sexual excitement is reached during the third phase. Involuntary muscle contractions, heightened blood pressure and heart rate, rapid intake of oxygen, sphincter muscle contraction, spasms of the carpopedel muscles in the feet, and sudden forceful release of sexual tension characterize the orgasmic phase.

For men, orgasm generally climaxes in the ejaculation of semen, which contains millions of sperm. Ejaculation consists of two steps. During the first phase, called the emission phase, seminal fluid builds up in the urethral bulb of the prostate gland. As the fluid accumulates, the male senses he is about to ejaculate. This is often experienced as inevitable and uncontrollable.

For women, orgasm also consists of rhythmic muscle contractions, in this case of the uterus, at about the same pace as in men. Tightening of the woman's muscles puts pressure on the man's penis and assists in male orgasm.

For both sexes, barring the presence of some form of sexual dysfunction, orgasm is an intensely pleasurable experience. Indeed, some see it as the most pleasurable experience possible.

THE FINAL PHASE OF SEXUAL RESPONSE

In this phase, the resolution, the body returns to normal levels of heart rate, blood pressure, breathing, and muscle contraction. Swelled and erect body parts return to normal and skin flushing disappears.

The resolution phase is marked by a general sense of well being and enhanced intimacy and possibly by fatigue as well. Many women are capable of a rapid return to the orgasmic phase with minimal stimulation and may experience continued orgasms for up to an hour.

Males, especially as they age, experience a refractory period of varying duration after orgasm. During this period, men cannot achieve orgasm, although partial or full erection may sometimes be maintained.

The duration of the refractory period can vary from just a few minutes to several days and there is great variability in the length of the refractory period both within and between men.

33 Sex is Not a Natural Act: "Historical, Scientific, Clinical, and Feminist Criticisms of the 'Human Sexual Response Cycle' Model"

Leonore Tiefer

The sexuality that is measured is taken to be the definition of sexuality itself.

—Lionel Trilling

THE HUMAN SEXUAL RESPONSE CYCLE METAPHOR: A UNIVERSAL MACHINE WITHOUT A MOTOR

The idea of the human sexual response cycle (HSRC) by that name was initially introduced by William Masters and Virginia Johnson (1966) to describe the sequence of physiological changes they observed and measured during laboratory-performed sexual activities such as masturbation and coitus. The goal of their research was to answer the question: "What physical reactions develop as the human male and female respond to effective sexual stimulation?" (Masters and Johnson, 1966, p. 4). Although they coined terms for their four stages, it appears that the metaphor of "the" overall sexual "cycle" was assumed from the very outset. They wrote: "A more concise picture of physiologic reaction to sexual stimuli may be presented by dividing *the human male's and female's cycles* of sexual response into four separate phases. This arbitrary four-part division of *the sexual response cycle* provides an effective framework for detailed description of physiological variants in sexual reaction" (p. 4, emphasis added).

The cycle metaphor indicates that Masters and Johnson envisioned sexual response from the start as a built-in, orderly sequence of events that would tend to repeat itself. The idea of a four-stage cycle brings to mind examples such as the four seasons of the annual calendar or the four-stroke internal combustion engine. Whether the cycle is designed by human agency or "nature," once begun it cycles independently of its origins, perhaps with some variability, but without reorganization or added stages, and the same cycle applies to everyone.

The idea of a sexual response cycle has some history, although its precursors focused heavily on an element omitted from the HSRC—the idea of sexual drive. In his intellectual history of modern sexology, Paul Robinson (1976) saw the origin of Masters and Johnson's four-stage HSRC in Havelock Ellis's theme of "tumescence and detumescence."

But the language of tumescence and detumescence was popular even prior to Ellis. In his analysis of Freud's theory of the libido, Frank Sulloway (1979) discussed nineteenth-century German and Austrian sexological ideas in circulation while Freud was writing. Sulloway pointed out that many sexological terms associated with Freud, such as *libido* and *erotogenic zones*, were in widespread use in European medical writings by the turn of the century, and he credited Albert Moll (then "possibly the best-known authority on sexual pathology in all of Europe" though "an obscure figure today") with originating a theory of two sexual drives—one of attraction and the other of detumescence (Sulloway, 1979, p. 302).

It is significant that, despite this long heritage of sexologic theorizing about sexual "energy," Masters and Johnson's model of sexual response did not include initiating components. Their omission of sexual drive, libido, desire, passion, and the like would return to haunt clinical sexology in the 1970s. Actually, in avoiding discussion of sexual drive, Masters and Johnson were following a trend peculiar to sexologists (in contrast to psychiatrists and psychoanalysts) during the twentieth century. Perhaps because of the history of elaborate but vague nineteenth-century writings, perhaps because of the subjective connotations to *desire*, talk of sex drive seemed to cause nothing but confusion for modern sexual scientists interested in operational definitions. Kinsey used the term only in passing, and meant by it "sexual capacity," the capacity to respond to stimulation with physical arousal (e.g., Kinsey, Pomeroy, Martin, and Gebhard, 1953, p. 102). Sexologists could compare individuals and groups in terms of this hypothetical internal mechanism, capacity, by looking at their frequencies of sexual behavior, thresholds for response, and so on with no reference to internal experience.

Frank Beach (1956), writing during the time Masters and Johnson were beginning their physiological observations, argued that talking about sex *drive* is usually circular and unproductive and approvingly noted that even Kinsey "equates sexual drive with frequency of orgasm." Beach suggested that sexual drive had nothing to do with "genuine biological or tissue needs" and that the concept should be replaced by sexual *appetite*, which is "a product of experience,...[with] little or no relation to biological or physiological needs" (Beach, 1956, p. 4). Although the concept of appetite never caught on in sexology, the recent rediscovery of "desire" indicates that ignoring the issue of initiation of sexual behaviors did not solve the problem.

By omitting the concept of drive from their model, Masters and Johnson eliminated an element of sexuality that is notoriously variable within populations and succeeded in proposing a universal model seemingly without much variability. In what I think is the only reference to sexual drive in their text, Masters and Johnson indicated their belief that the sexual response cycle was actually an inborn drive to orgasm: "The cycle of sexual response, with orgasm as the ultimate point in progression, generally is believed to develop from a drive of biologic origin deeply integrated into the condition of human existence" (Masters and Johnson, 1966, p. 127). The cycle of sexual response, then, reflects the operation of an inborn program, like the workings of a mechanical clock. As long as the "effective sexual stimulation" (i.e., energy source) continues, the cycle proceeds through its set sequence.

SCIENTIFIC CRITICISMS OF THE HSRC MODEL

Masters and Johnson proposed a universal model for sexual response. At no point did they talk of "a" human sexual response cycle, but only of "the" human sexual response cycle. The critique of the HSRC model begins with a discussion of the generalizability of Masters and Johnson's research results. Analysis of their work shows that the existence of the HSRC was assumed before the research began and that this assumption guided subject selection and research methods.

Subject Selection Biases: Orgasm with Coital and Masturbatory Experience

In a passage buried four pages from the end of their text, Masters and Johnson revealed that for their research they had established "a requirement that there be a positive history of masturbatory and coital orgasmic experience before any study subject [could be] accepted into the program" (Masters and Johnson, 1966, p. 311, emphasis added). This requirement in and of itself would seem to invalidate any notion that the HSRC is universal. It indicates that Masters and Johnson's research was designed to identify physiological functions of subjects who had experienced *particular*, preselected sexual responses. That is, rather than the HSRC being the best-fit model chosen to accommodate the results of their research, the HSRC actually guided the selection of subjects for the research.

Two popularizations of Masters and Johnson's physiological research commented on this element of subject selection but disregarded its implications for HSRC generalizability:

> Men and women unable to respond sexually and to reach orgasm were also weeded out. Since this was to be a study of sexual responses, those unable to respond could contribute little to it. (Brecher and Brecher, 1966, p. 54)

> If you are going to find out what happens, obviously you must work with those to whom it happens. (Lehrman, 1970, p. 170)

"Unable to respond"? If you want to study human singing behaviors, do you only select international recording artists? One could just as easily argue that there are many sexually active and sexually responsive men and women who do not regularly experience orgasm during masturbation and/or coitus whose patterns of physiological arousal and subjective pleasure were deliberately excluded from the sample. No research was undertaken to investigate "human" sexual physiology and subjectivity, only to measure the responses of an easily orgasmic sample. The "discovery" of the HSRC was a self-fulfilling prophecy, with the research subjects selected so as to compress diversity. The HSRC cannot be universalized to the general population.

The apparently identical performance requirements for male and female research subjects masked the bias of real-world gender differences in masturbatory experience. Masters and Johnson began their physiological research in 1954. In 1953, the Kinsey group had reported that only "58 percent of the females" in their sample had been "masturbating to orgasm at some time in their lives" (Kinsey, Pomeroy, Martin, and Gebhard, 1953, p. 143). Married women, the predominant subjects in Masters and Johnson's research, had even lower masturbatory frequencies than divorced or single women. This contrasts with the 92 percent incidence of men with masturbatory experience reported by the same researchers (Kinsey, Pomeroy, and Martin, 1948, p. 339). Masters and Johnson had to find men and women with similar sexual patterns despite having been raised in dissimilar sociosexual worlds. Obviously, because of this requirement the women research participants were less representative than the men.

Subject Selection Biases: Class Differences

Just as Masters and Johnson chose subjects with certain types of sexual experiences, they deliberately chose subjects who did not represent a cross-section of socioeconomic backgrounds. They wrote: "As discussed, the sample was weighted *purposely* toward higher than average intelligence levels and socioeconomic backgrounds. *Further selectivity* was established ... to determine willingness to participate, facility of sexual responsiveness, and ability to communicate finite details of sexual reaction" (Masters and Johnson, 1966, p. 12, emphasis added). Masters and Johnson's popularizers disparaged the possible bias introduced by this selectivity with such comments as, "The higher than average educational level of the women volunteers is hardly likely to affect the acidity of their vaginal fluids" (Brecher and Brecher, 1966, p. 60).

But one cannot simply dismiss possible class differences in physiology with an assertion that there are none. *Could* differences in social location affect the physiology of sexuality? The irony of assuming that physiology is universal and therefore that class differences make no difference is that no one conducts research that asks the question.

In fact, Kinsey and his colleagues had shown wide differences between members (especially males) of different socioeconomic classes with regard to incidence and prevalence of masturbation, premarital sexual activities, petting (including breast stimulation), sex with prostitutes, positions used in intercourse, oral-genital sex, and even nocturnal emissions. For example, "There are 10 to 12 times as frequent nocturnal emissions among males of the upper educational classes as there are among males of the lower classes" (Kinsey, Pomeroy, and Martin, 1948, p. 345). Kinsey noted, "It is particularly interesting to find that there are [great] differences between educational levels in regard to nocturnal emissions—a type of sexual outlet which one might suppose would represent involuntary behavior" (p. 343). Given this finding, doesn't it seem possible, even likely, that numerous physiological details might indeed relate to differences in sexual habits? Kinsey also mentioned class differences in latency to male orgasm (p. 580). The more the variation in physiological details among subjects from different socio-economic backgrounds, the less the HSRC is appropriate as a universal norm.

Subject Selection Biases: Sexual Enthusiasm

Masters and Johnson concluded their physiological research text as follows: "Through the years of research exposure, the one factor in sexuality that consistently has been present among members of the study-subject population has been a basic interest in and desire for effectiveness of sexual performance. *This one factor may represent the major area of difference between the research study subjects and the general population*" (Masters and Johnson, 1966, p. 315, emphasis added).

Masters and Johnson do not explain what they mean by their comment that "the general population" might not share the enthusiasm for sexual performance of their research subjects and do not speculate at all on the possible impact of this comment on the generalizability of their results. Whereas at first it may seem reasonable to assume that everyone has "a basic interest in and desire for *effectiveness* of sexual *performance*," on closer examination the phrase "effectiveness of sexual performance" seems not so much to characterize everyone as to identify devotees of a particular sexual style.

We get some small idea of Masters and Johnson's research subjects from the four profiles given in Chapter 19 of *Human Sexual Response* (1966). These profiled subjects were selected by the authors from the 382 women and 312 men who participated in their study. The two women described had masturbated regularly (beginning at ages ten and fifteen, respectively), had begun having intercourse in adolescence (at ages fifteen and seventeen), and were almost always orgasmic and occasionally multiorgasmic in the lab-oratory. For the first woman, twenty-six and currently unmarried, it was explicitly stated that "sexual activity [was] a major factor in [her] life" (Masters and Johnson, 1966, p. 304) and that she became a research subject because of "financial demand and sexual tension" (p. 305). No comparable information was given about the second woman, who was thirty-one and married, but she and her husband had "stated categorically" that they had "found [research participation] of significant importance in their marriage" (p. 307).

The unmarried male subject, age twenty-seven, was described as having had adolescent onset of masturbation, petting, and heterosexual intercourse as well as four reported homosexual experiences at different points in his life. The married man, age thirty-four, had had little sexual experience until age twenty-five. He and his wife of six years had joined the research program "hoping to acquire knowledge to enhance the sexual component of their marriage" (Masters and Johnson, 1966, p. 311). The researchers noted, "[His] wife has stated repeatedly that subsequent to [research project] participation her husband has been infinitely more effective both in stimulating and satisfying her sexual tensions. He in turn holds her sexually responsive without reservation. Her freedom and security of response are particularly pleasing to him" (p. 311).

Every discussion of sex research methodology emphasizes the effects of volunteer bias and bemoans the reliance on samples of convenience that characterizes its research literature (e.g., Green and Wiener, 1980). Masters and Johnson make no attempt to compare their research subjects with any other research sample, saying, "There are no established norms for male and female sexuality in our society ... [and] there is no scale with which to measure or evaluate the sexuality of the male and female study-subject population" (Masters and Johnson, 1966, p. 302). Although there may not be "norms," there are other sex research surveys of attitudes and behavior. For example, volunteers for sex research are usually shown to be more liberal in their attitudes than socioeconomically comparable nonvolunteer groups (Hoch, Safir, Peres, and Shepher, 1981; Clement, 1990).

How might the sample's interest in "effective sexual performance" have affected Masters and Johnson's research and their description of the HSRC? The answer relates both to the consequences of ego-investment in sexual performance and to the impact of specialization in a sexual style focused on orgasm, and we don't know what such consequences might be. I cannot specify the effect of this sexually skewed sample any more than I could guess what might be the consequences for research on singing of only studying stars of the Metropolitan Opera. The point is that the subject group was exceptional, and only by *assuming* HSRC universality can we generalize its results to others.

Experimenter Bias in the Sexuality Laboratory

Masters and Johnson made no secret of the fact that subjects volunteering for their research underwent a period of adjustment, or a "controlled orientation program," as they called it (Masters and Johnson, 1966, p. 22). This "period of training" helped the subjects "gain confidence in their ability to respond successfully while subjected to a variety of recording devices" (p. 23). Such a training period provided an opportunity for numerous kinds of "experimenter biases," as they are known in social psychology research, wherein the expectations of the experimenters are communicated to the subjects and have an effect on their behavior (Rosenthal, 1966). The fact that Masters and Johnson repeatedly referred to episodes of sexual activity with orgasm as "successes" and those without orgasm or without rigid erection or rapid ejaculation as "failures" (e.g., Masters and Johnson, 1966, p. 313) makes it seem highly likely that their performance standards were communicated to their subjects. Moreover, they were candid about their role as sex therapists for their subjects: "When female orgasmic or male ejaculatory failures develop in the laboratory, the *situation is discussed* immediately. Once the individual has been *reassured, suggestions* are made for improvement of future performance" (p. 314, emphasis added).

Another example of the tutelage provided is given in the quotation from the thirty-four-year-old man described in Chapter 19 of their book. He and his wife had entered the program hoping to obtain sexual instruction and seemed to have received all they expected and more. Masters and Johnson appeared to be unaware of any incompatibility between the roles of research subject and student or patient. Again, this reveals their preexisting standards for sexual response and their interest in measuring in the laboratory only sexual patterning consisting of erections, orgasms, ejaculations, whole-body physical arousal, and so on, that is, that which they already defined as sexual response.

In addition to overt instruction and feedback, social psychology alerts us to the role of covert cues. Research has shown that volunteer subjects often are more sensitive to experimenters' covert cues than are non-volunteers (Rosenthal and Rosnow, 1969). One could speculate that sex research volunteers characterized by a "desire for effective sexual performance" may well be especially attentive to covert as well as overt indications that they are performing as expected in the eyes of the white-coated researchers.

The Bias of "Effective" Sexual Stimulation

As mentioned near the beginning of this chapter, Masters and Johnson set out to answer the question, "What physical reactions develop as the human male and female respond to effective sexual stimulation?" (Masters and Johnson, 1966, p. 4). What is "effective" sexual stimulation? In fact, I think this is a key question in deconstructing the HSRC. Masters and Johnson stated, "It constantly should be borne in mind that the primary research interest has been concentrated quite literally upon what men and women do in response to effective sexual stimulation" (p. 20).

The *intended* emphasis in this sentence, I believe, is that the authors' "primary" interest was not in euphemism, and not in vague generality, but in the "literal" physical reactions people experience during sexual activity. I think the *actual* emphasis of the sentence, however, is that the authors were interested in only one type of sexual response, that which people experience in reaction to a particular type of stimulation. Such a perspective would be akin to vision researchers only being interested in optic system responses to lights of certain wavelengths, say, red and yellow, or movement physiologists only being interested in physical function during certain activities, such as running.

In each of the book's chapters devoted to the physical reactions of a particular organ or group of organs (e.g., clitoris, penis, uterus, respiratory system), Masters and Johnson began by stating their intention to look at the responses to "effective sexual stimulation." But where is that specific type of stimulation described? Although the phrase appears dozens of times in the text, it is not in the glossary or the index, and no definition or description can be found. The reader must discover that *"effective sexual stimulation" is that stimulation which facilitates a response that conforms to the HSRC.* This conclusion is inferred from observations such as the following, taken from the section on labia minora responses in the chapter on "female external genitalia": "Many women have progressed well into plateau-phase levels of sexual response, had the effective stimulative techniques withdrawn, and been unable to achieve orgasmic-phase tension release.... When an obviously effective means of sexual stimulation is withdrawn and orgasmic-phase release is not achieved, the minor-labial coloration will fade rapidly" (Mas ters and Johnson, 1966, p. 41).

Effective stimulation is that stimulation which facilitates "progress" from one stage of the HSRC to the next, particularly that which facilitates orgasm. Any stimulation resulting in responses other than greater physiological excitation and orgasm is defined by exclusion as "ineffective" and is not of interest to these authors.

This emphasis on "effective stimulation" sets up a tautology comparable to that resulting from biased subject selection. The HSRC cannot be a scientific *discovery* if the acknowledged "primary research interest" was to study stimulation defined as that which facilitates the HSRC. Again, the HSRC, "with orgasm as the ultimate point in progression" (Masters and Johnson, 1966, p. 127), preordained the results.

CLINICAL CRITICISMS OF THE HSRC MODEL

The HSRC model has had a profound impact on clinical sexology through its role as the centerpiece of contemporary diagnostic nomenclature. In this section, I will first discuss how contemporary nomenclature came to rely on the HSRC model and then describe what I see as several deleterious consequences.

HSRC and the DSM Classification of Sexual Disorders

I have elsewhere detailed the development of sexual dysfunction nosology in the four sequential editions of the American Psychiatric Association's *Diagnostic and Statistical Manual of Mental Disorders* (*DSM*) (Tiefer, 1992b). Over a period of thirty-five years, the nosology evolved from not listing sexual dysfunctions at all (APA, 1952, or DSM-I) to listing them as symptoms of psychosomatic disorders (APA, 1968, or DSM-II), as a subcategory of psychosexual disorders (APA, 1980, or DSM-III), and as a subcategory of sexual disorders (APA, 1987, or DSM-III-R).

The relation of this nosology to the HSRC language can be seen in the introduction to the section on sexual dysfunctions (identical in both DSM-III and DSM-III-R)

> The *essential feature* is inhibition in the appetitive or psychophysiological changes that characterize *the complete sexual response cycle*. The complete sexual response cycle can be divided into the following phases: 1. Appetitive. This consists of fantasies about sexual activity and a desire to have sexual activity. 2. Excitement. This consists of a subjective sense of sexual pleasure and accompanying physiological changes. ... 3. Orgasm. This consists of a peaking of sexual pleasure, with release of sexual tension and rhythmic contraction of the perineal muscles and pelvic reproductive organs. ... 4. Resolution. This consists of a sense of general relaxation, well-being, and muscular relaxation. (APA, 1987, pp. 290-291, emphasis added)

In fact, this cycle is not identical to Masters and Johnson's HSRC (although it, too, uses the universalizing language of "the" sexual response cycle). The first, or appetitive, phase was added when sexologists confronted clinical problems having to do with sexual disinterest. In their second book (1970), Masters and Johnson loosely used their HSRC physiological research to generate a list of sexual dysfunctions: premature ejaculation, ejaculatory incompetence, orgasmic dysfunction (women's), vaginismus, and dyspareunia (men's and women's). These were put forth as deviations from the HSRC that research had revealed as the norm. By the late 1970s, however, clinicians were describing a syndrome of sexual disinterest that did not fit into the accepted response cycle. Helen Singer Kaplan argued that a "separate phase [sexual desire] which had previously been neglected, must be added for conceptual completeness and clinical effectiveness" (Kaplan, 1979, p. xviii). *DSM-III* and *DSM-III-R* then merged the original HSRC with the norm of sexual desire to generate "the complete response cycle" presented above.

Clearly, the idea and much of the language of the nosology derived from Masters and Johnson's work, and in fact they are cited in the *DSM* footnotes as the primary source. Is it appropriate to use the HSRC to generate a clinical standard of normality? Is it appropriate to enshrine the HSRC as the standard of human sexuality such that deviations from it become the essential feature of abnormality?

Let us briefly examine how sexual problems are linked to mental disorders in the *DSM* and how the HSRC was used in the sexuality section. The definition of mental disorder offered in *DSM-III* specifies:

> In DSM-III each of the mental disorders is conceptualized as a clinically significant behavioral or psychological syndrome or pattern that occurs in an individual and that is typically associated with either a painful symptom (distress) or impairment in one or more areas of function (disability). In addition, there is an inference that there is a behavioral, psychological or biological dysfunction. (APA, 1980, p. 6)

In an article introducing the new classification scheme to the psychiatric profession, the APA task force explained their decisions. With regard to sexual dysfunctions, the task force members had concluded that "inability to experience *the normative sexual response cycle* [emphasis added] represented a *disability* in *the important area* of sexual functioning, whether or not the individual was distressed by the symptom" (Spitzer, Williams, and Skodol, 1980, pp. 153-154). That is, deviation from the now-normative sexual response cycle was to be considered a disorder even if the person had no complaints.

The diagnostic classification system clearly assumed that the HSRC was a universal bedrock of sexuality. Yet I have shown that it was a self-fulfilling result of Masters and Johnson's methodological decisions rather than a scientific discovery. It was the result of a priori assumptions rather than empirical research. Arguably, a clinical standard requires a greater demonstration of health impact and universal applicability than that offered by Masters and Johnson's research.

In fact, it is likely the case that the *DSM* authors adopted the HSRC model because it was useful and convenient. Professional and political factors that probably facilitated the adoption include professional needs within psychiatry to move away from a neurosis disorder model to a more concrete and empirical model, legitimacy needs within the new specialty of sex therapy, and the interests of feminists in progressive sexual standards for women (Tiefer, 1992b). Thus, the enshrinement of the HSRC and its upgraded versions as the centerpiece of sexual dysfunction nomenclature in *DSM-III* and *DSM-III-R* is not scientifically reliable and represents a triumph of politics and professionalism.

Sexuality as the Performances of Fragmented Body Parts

One deleterious clinical consequence of the utilization of the HSRC model as the sexual norm has been increased focus on segmented psychophysiological functioning. Just for example, consider the following disorder descriptions, which appear in *DSM-III-R*:

1. "partial or complete failure to attain or maintain the lubrication-swelling response of sexual excitement [Female Arousal Disorder]"
2. "involuntary spasm of the musculature of the outer third of the vagina that interferes with coitus [Vaginismus]"
3. "inability to reach orgasm in the vagina [Inhibited Male Orgasm]"

In the current nosology, the body as a whole is never mentioned but in stead has become a fragmented collection of parts that pop in and out at different points in the performance sequence. This compartmentalization lends itself to mechanical imagery, to framing sexuality as the smooth operation and integration of complex machines, and to seeing problems of sexuality as "machines in disrepair" that need to be evaluated by high-technology part-healers (Soble, 1987). If there is a sexual problem, check each component systematically to detect the component out of commission. Overall satisfaction (which is mentioned nowhere in the nosology) is assumed to be a result of perfect parts-functioning. Recall that subjective distress is not even required for diagnosis, just objective indication of deviation from the HSRC.

This model promotes the idea that sexual disorder can be defined as deviation from "normal" as indicated by medical test results. A bit of thought, however, will show that identifying proper norms for these types of measurements is a tricky matter. How rigid is rigid? How quick is premature? How delayed is delayed? The answers to these questions are more a product of expectations, cultural standards, and particular partner than they are of objective measurement. And yet a series of complex and often invasive genital measurements are already being routinely used in evaluations of erectile dysfunction (Krane, Goldstein, and DeTejada, 1989). Norms for many of the tests are more often provided by medical technology manufacturers than by scientific research, and measurements on nonpatient samples are often lacking. Despite calls for caution in use and interpretation, the use of sexuality measurement technology continues to escalate (Burris, Banks, and Sherins, 1989; Kirkeby, Andersen, and Poulson, 1989; Schiavi, 1988; Sharlip, 1989).

This example illustrates a general medical trend: while reliance on tests and technology for objective information is increasing, reliance on patients' individualized standards and subjective reports of illness is decreasing (Osherson and AmaraSingham, 1981). The end result may be, as Lionel Trilling (1950) worried in a review of the first Kinsey report, that "the sexuality that is measured is taken to be the definition of sexuality itself (p. 223)." Although it seems only common sense and good clinical practice to want to "rule out" medical causes prior to initiating a course of psychotherapeutic or couple treatment for sexual complaints, such "ruling out" has become a growth industry rather than an adjunct to psychological and couple-oriented history-taking. Moreover, there is a growing risk of iatrogenic disorders being induced during the extensive "ruling out" procedures.

The HSRC has contributed significantly to the idea of sexuality as proper parts-functioning. Masters and Johnson's original research can hardly be faulted for studying individual physiological components to answer the question, "What physical reactions develop as the human male and female respond to effective sexual stimulation?" But once the physiological aspects became solidified into a universal, normative sequence known as "the" HSRC, the stage was set for clinical preoccupation with parts-functioning. Despite Masters and Johnson's avowed interest in sexuality as communication, intimacy, self-expression, and mutual pleasuring, their clinical ideas were ultimately mechanical (Masters and Johnson, 1975).

Exclusive Genital (i.e., Reproductive) Focus for Sexuality

"Hypoactive sexual desire" is the only sexual dysfunction in the DSM-III-R defined without regard to the genital organs. "Sexual aversion," for example, is specifically identified as aversion *to the genitals*. The other sexual dysfunctions are defined in terms *genital* pain, spasm, dryness, deflation, uncontrolled responses, delayed responses, too-brief responses, or absent responses. The DSM locates the boundary between normal and abnormal (or between healthy and unhealthy) sexual function exclusively on genital performances.

"Genitals" are those organs involved in acts of generation, or biological reproduction. Although the DSM does not explicitly endorse reproduction as the primary purpose of sexual activity, the genital focus of the sexual dysfunction nosology implies such a priority. The only sexual acts mentioned are coitus, [vaginal] penetration, sexual intercourse, and noncoitalditoral stimulation. Only one is not a heterosexual coital act. Masturbation is only mentioned as a "form of stimulation." Full *genital performance during heterosexual intercourse is the essence of sexual functioning*, which excludes and demotes nongenital possibilities for pleasure and expression. Involvement or noninvolvement of the nongenital body becomes incidental, of interest only as it impacts on genital responses identified in the nosology.

Actually, the HSRC is a whole-body response, and Masters and Johnson were as interested in the physiology of "extragenital" responses as genital ones. Yet the stages of the HSRC as reflected in heart rate or breast changes did not make it into the DSMs. As Masters and Johnson transformed their physiological cycle into a clinical cycle, they privileged a reproductive purpose for sexuality by focusing on the genitals. It would seem that once they turned their interest to sexual problems rather than sexual process, their focus shifted to *sexuality as outcome*.

There is no section on diagnosis in Masters and Johnson's second clinical book (1970), no definition of normal sexuality, and no hint of how the particular list of erectile, orgasmic, and other genitally focused disorders was derived. The authors merely described their treatments of "the specific varieties of sexual dysfunction that serve as presenting complaints of patients referred" (Masters and Johnson, 1970, p. 91). But surely this explanation cannot be the whole story. Why did they exclude problems like "inability to relax,...attraction to partner other than mate,...partner chooses inconvenient time, ... too little tenderness" or others of the sort later labeled "sexual difficulties" (Frank, Anderson, and Rubinstein, 1978)? Why did they exclude problems like "partner is only interested in orgasm,...partner can't kiss,...partner is too hasty,...partner has no sense of romance," or others of the sort identified in surveys of women (Hite, 1976)?

In fact, the list of disorders proposed by Masters and Johnson seems like a list devised by Freudians who, based on their developmental stage theory of sexuality, define genital sexuality as the sine qua non of sexual maturity. Despite the whole-body focus of the HSRC physiology research, the clinical interest of its authors in proper genital performance as the essence of normal sexuality indicates their adherence to an earlier tradition. The vast spectrum of sexual possibility is narrowed to genital, that is, to reproductive performance.

Symptom Reversal as the Measure of Sex Therapy Success

A final undesirable clinical consequence of the HSRC and its evolution in the DSM is the limitation it imposes on the evaluation of therapy success. Once sexual disturbances are defined as specific malperformances within "the" sexual response cycle, evaluation of treatment effectiveness narrows to symptom reversal.

But the use of symptom reversal as the major or only measure of success contrasts with sex therapy as actually taught and practiced (Hawton, 1985). Typical practice focuses on individual and relationship satisfaction and includes elements such as education, permission-giving, attitude change, anxiety reduction, improved communication, and intervention in destructive sex roles and lifestyles (LoPiccolo, 1977). A recent extensive survey of 289 sex therapy providers in private practice reinforced the statement that "much of sex therapy actually was nonsexual in nature" and confirmed that therapy focuses on communication skills, individual issues, and the "nonsexual relationship" (Kilmann et al., 1986).

Follow-up studies measuring satisfaction with therapy and changes in sexual, psychological, and interpersonal issues show varying patterns of improvement, perhaps because therapists tend to heedlessly lump together cases with the "same" symptom. It is erroneous to assume that couples and their experience of sex therapy are at all homogeneous, despite their assignment to specific and discrete diagnostic categories based on the HSRC. Citing his own "painful experience" (Bancroft, 1989, p. 489) with unreplicable results of studies comparing different forms of treatment, John Bancroft suggested that there is significant prognostic variability among individuals and couples even within diagnostic categories. He concluded, "It may be that there is no alternative to defining various aspects of the sexual relationship, e.g., sexual response, communication, enjoyment, etc. and assessing each separately" (p. 497).

It might be thought that using symptom reversal as the measure of success is easier than evaluating multiple issues of relationship satisfaction, but this is not true, since *any* measure of human satisfaction needs to be subtle. That is, it is indeed easy to measure "success" with objective technologies that evaluate whether a prosthesis successfully inflates or an injection successfully produces erectile rigidity of a certain degree. When evaluating the human success of physical treatments, however, researchers invariably introduce complex subjective elements. The questions they select, the way they ask the questions, and their interpretations of the answers are all subjective (Tiefer, Pedersen, and Melman, 1988). In evaluating patients' satisfaction with penile implant treatment, asking the patients whether they would have prosthesis surgery again produces different results from evaluating postoperative satisfaction with sexual frequency, the internal feeling of the prosthesis during sex, anxieties about the indwelling prosthesis, changes in relationship quality, and so on.

The present diagnostic nomenclature, based on the genitally focused HSRC, results in evaluation of treatment success exclusively in terms of symptom reversal and ignores the complex sociopsychological context of sexual performance and experience. The neat four-stage model, the seemingly clean clinical typology, all result in neat and clean evaluation research—which turns out to relate only partially to real people's experiences.

FEMINIST CRITICISMS OF THE HSRC MODEL

Paul Robinson (1976) and Janice Irvine (1990) have discussed at length how Masters and Johnson deliberately made choices throughout *Human Sexual Response* and *Human Sexual Inadequacy* to emphasize male-female sexual similarities. The most fundamental similarity, of course, was that men and women had identical HSRCs. The diagnostic nomenclature continues this emphasis by basing the whole idea of sexual dysfunction on the gender-neutral HSRC and by scrupulously assigning equal numbers and parallel dysfunctions to men and women. (Desire disorders are not specified as to gender; other dysfunctions include one arousal disorder for each gender, one inhibited orgasm disorder for each gender, premature ejaculation for men and vaginismus for women, and dyspareunia, which is defined as "recurrent or persistent genital pain in either a male or a female.")

Yet, is the HSRC really gender-neutral? Along with other feminists, I have argued that the HSRC model of sexuality, and its elaboration and application in clinical work, favors men's sexual interests over those of women (e.g., Tiefer, 1988a). Some have argued that sex role socialization introduces fundamental gender differences and inequalities into adult sexual experience that cannot be set aside by a model that simply proclaims male and female sexuality as fundamentally the same (Stock, 1984). I have argued that the HSRC, with its alleged gender equity, disguises and trivializes *social* reality, that is, gender inequality (Tiefer, 1990a) and thus makes it all the harder for women to become sexually equal in fact.

Let's look briefly at some of these gender differences in the real world. First, to oversimplify many cultural variations on this theme, men and women are raised with different sets of sexual values—men toward varied experience and physical gratification, women toward intimacy and emotional communion (Gagnon, 1977; Gagnon, 1979; Gagnon and Simon, 1969; Simon and Gagnon, 1986). By focusing on the physical aspects of sexuality and ignoring the rest, the HSRC favors men's value training over women's. Second, men's greater experience with masturbation encourages them toward a genital focus in sexuality, whereas women learn to avoid acting on genital urges because of the threat of lost social respect. With its genital focus, the HSRC favors men's training over women's. As has been mentioned earlier, by requiring experience and comfort with masturbation to orgasm as a criterion for all participants, the selection of research subjects for *Human Sexual Response* looked gender-neutral but, in fact, led to an unrepresentative sampling of women participants.

Third, the whole issue of "effective sexual stimulation" needs to be addressed from a feminist perspective. As we have seen, the HSRC model was based on a particular kind of sexual activity, that with "effective sexual stimulation." Socioeconomic subordination, threats of pregnancy, fear of male violence, and society's double standard reduce women's power in heterosexual relationships and militate against women's sexual knowledge, sexual assertiveness, and sexual candor (Snitow, Stansell, and Thompson, 1983; Vance, 1984). Under such circumstances, it seems likely that "effective sexual stimulation" in the laboratory or at home favors what men prefer.

The HSRC assumes that men and women have and want the same kind of sexuality since physiological research suggests that in some ways, and under selected test conditions, we are built the same. Yet social realities dictate that we are not all the same sexually-not in our socially shaped wishes, in our sexual self development, or in our interpersonal sexual meanings. Many different studies-from questionnaires distributed by feminist organizations to interviews of self-defined happily married couples, from popular magazine surveys to social psychologists' meta-analyses of relationship research-show that women rate affection and emotional communication as more important than orgasm in a sexual relationship (Hite, 1987; Frank, Anderson, and Rubinstein, 1978; Tavris and Sadd, 1977; Peplau and Gordon, 1985). Given this evidence, it denies women's voices entirely to continue to insist that sexuality is best represented by the universal "cycle of sexual response, with orgasm as the ultimate point in progression" (Masters and Johnson, 1966, p. 127).

Masters and Johnson's comparisons of the sexual techniques used by heterosexual and homosexual couples can be seen to support the claim that "effective sexual stimulation" simply means what men prefer. Here are examples of the contrasts:

> The sexual behavior of the married couples was far more performance-oriented...Preoccupation with orgasmic attainment was expressed time and again by heterosexual men and women during interrogation after each testing session...[By contrast] the committed homosexual couples took their time in sexual interaction in the laboratory...In committed heterosexual couples' interaction, the male's sexual approach to the female,...rarely more than 30 seconds to a minute, were spent holding close or caressing the total body area before the breasts and/or genitals were directly stimulated. This was considerably shorter than the corresponding time interval observed in homosexual couples. (Masters and Johnoson, 1979, 99. 64-65, 66)

After describing various techniques of breast stimulation, the authors reported that heterosexual women enjoyed it much less than lesbian but that "all the [heterosexual] women thought that breast play was very important in their husband's arousal" (p.67). The authors repeatedly emphasized that the differences between lesbians and heterosexual techniques were greater than between heterosexual and male homosexual techniques.

The enshrinement of the HSRC in the *DSM* diagnostic nomenclature represented the ultimate in context-stripping, as far as women's sexuality is concerned. To speak merely of desire, arousal, and orgasm as constitutive of sexuality and ignore relationships and women's psychosocial development is to ignore women's experiences of exploitation, harassment, and abuse and to deny women's social limitations. To reduce sexuality to the biological specifically disadvantages women, feminists argue, because women as a class are disadvantaged by social sexual reality (Laws, 1990; Hubbard, 1990; Birke, 1986).

Finally, the biological reductionism of the HSRC and the DSM is subtly conveyed by their persistent use of the terms *males* and *females* rather than *men* and *women*. There are no men and no women in the latest edition of the diagnostic nomenclature, only male and females and vaginas and so forth. In *Human Sexual Response*, men and women appear in the text from time to time, but only males and females make it to the chapter headings, and a rough count of a few pages here and there in the text reveals a 7:1 use of the general animal kingdom terms over the specifically human ones. A feminist deconstruction of the HSRC and of contemporary perspectives on sexuality could do worse than begin by noticing and interpreting how the choice of vocabulary signals the intention to ignore culture.

CONCLUSION

I have argued in this chapter that the human sexual response cycle (HSRC) model of sexuality is flawed from scientific, clinical, and feminist points of view. Popularized primarily because clinicians and researchers needed norms that were both objective and universal, the model is actually neither objective nor universal. It imposes a false biological uniformity on sexuality that does not support the human uses and meanings of sexual potential. The most exciting work in sex therapy evolves toward systems analyses and interventions that combine psychophysiological sophistication with respect for individual and couple diversity (e.g., Verhulst and Heiman, 1988). Subjective dissatisfactions are seen more as relative dyssynchronies between individuals or between elements of culturally based sexual scripts than as malfunctions of some universal sexual essence.

Defining the essence of sexuality as a specific sequence of physiological changes promotes biological reductionism. Biological reductionism not only separates genital sexual performance from personalities, relationships, conduct, context, and values but also overvalues the former at the expense of the latter. As Abraham Maslow (1966) emphasized, studying parts may be easier than studying people, but what do you understand when you're through? Deconstruction and desacralizing the HSRC should help sex research unhook itself from the albatross of biological reductionism.

34 Sex is Not a Natural Act: "The Kiss"

Leonore Tiefer

Nothing seems more natural than a kiss. Consider the French kiss, also known as the soul kiss, deep kiss, or tongue kiss (to the French it was the Italian kiss). Western societies regard this passionate exploration of mouths and tongues as an instinctive way to express love and to arouse desire. To a European who associates deep kisses with erotic response, the idea of one without the other feels like summer without sun.

Yet soul kissing is completely absent in many cultures of the world, where sexual arousal may be evoked by affectionate bites or stinging slaps. Anthropology and history amply demonstrate that, depending on time and place, the kiss may or may not be regarded as a sexual act, a sign of friendship, a gesture of respect, a health threat, a ceremonial celebration, or a disgusting behavior that deserves condemnation.

Considering the diversity of kissing customs, it astonishes me that so few social scientists have given the kiss any attention. Kissing is usually relegated to an occasional footnote, if authors bother to mention it at all. My computer search of kissing references in *Psychological Abstracts* and the index *Medicus* turned up some papers on mononucleosis ("the kissing disease"), one article on a fish known as the "kissing gourami," and unrelated work by people with names like Kissing and Kissler. Even sex researchers are uninterested. Sex now refers to intercourse, not kissing or petting. In textbooks on human sexuality, kissing rarely appears in the index.

ANTHROPOLOGY AND KISSING

I became fascinated by the remarkable cultural and historical variations in styles and purposes of kissing, given how "natural" it seems to pursue whatever customs each of us has grown accustomed to. Clellan Ford and Frank Beach (1951) compared the sexual customs of the many tribal societies that were recorded in the Human Relations Area Files at Yale. Few field studies mentioned kissing customs at all. Of the twenty-one that did, some sort of kissing accompanied intercourse in thirteen tribes—the Chiricahua, Cree, Crow, Gros

Ventre, Hopi, Huichol, Kwakiutl, and Tarahumara of North America; the Alorese, Keraki, Trobrianders, and Trukese of Oceania; and the Lapps in Eurasia. There were some interesting variations: the Kwakiutl, Trobrianders, Alorese, and Trukese kiss by sucking the lips and tongue of their partners; the Lapps like to kiss the mouth and nose at the same time.

But sexual kissing is unknown in many societies, including the Balinese, Chamorro, Manus, and Tinguian of Oceania; the Chewa and Thonga of Africa; the Siriono of South America; and the Lepcha of Eurasia. In such cultures, the mouth-to-mouth kiss is considered dangerous, unhealthy, or disgusting, the way Westerners might regard a custom of sticking one's tongue into a lover's nose. Ford and Beach reported that when the Thonga first saw Europeans kissing, they laughed, remarking, "Look at them—they eat each other's saliva and dirt."

Deep kissing apparently has nothing to do with degree of sexual inhibition or repression in a culture. On certain Polynesian islands, women are orgasmic and sexually active, yet kissing was unknown until Westerners and their popular films arrived. Research in parts of Ireland, by contrast, where sex was considered dirty and sinful, and, for women, a duty, shows that the Irish also were oblivious to tongue kissing until recent decades.

Many tribes across Africa and elsewhere believe that the soul enters and leaves through the mouth and that a person's bodily products can be collected and saved by an enemy for harmful purposes. In these societies, the possible loss of saliva would cause a kiss to be regarded as a dangerous gesture. There, the "soul kiss" is taken literally. (It was taken figuratively in Western societies; recall Christopher Marlowe's "Sweet Helen, make me immortal with a kiss! Her lips suck forth my soul.")

Although the deep kiss is relatively rare around the world as a part of sexual intimacy, other forms of mouth or nose contact are common—particularly the "oceanic kiss," named for its prevalence among cultures in Oceania but not limited to them. The Tinguians place their lips near the partner's face and suddenly inhale. Balinese lovers bring their faces close enough to catch each other's perfume and to feel the warmth of the skin, making contact as they move their heads slightly. Another kiss, as practiced by Chinese Yakuts and Mongolians at the turn of the century, has one person's nose pressed to the other's cheek, followed by a nasal inhalation and finally a smacking of lips.

The oceanic kiss may be varied by the placement of the nose and cheek, the vigor of the inhalation, the nature of the accompanying sounds, the action of the arms, and so on; it is used for affectionate greeting as well as for sexual play. Some observers think that the so-called Eskimo or Malay kiss of rubbing noses is actually a mislabeled oceanic kiss; the kisser is simply moving his or her nose rapidly from one cheek to the other of the partner, bumping noses en route.

Small tribes and obscure Irish islanders are not the only groups to eschew tongue kissing. The advanced civilizations of China and Japan, which regarded sexual proficiency as high art, apparently cared little about it. In their voluminous display of erotica—graphic depictions of every possible sexual position, angle of intercourse, and variation of partner and setting—mouth-to-mouth kissing is conspicuous by its absence. Japanese poets have rhapsodized for centuries about the allure of the nape of the neck, but they have been silent on the mouth;

indeed, kissing is acceptable only between mother and child. (The Japanese have no word for kissing—though they recently borrowed from English to create *kissu.*) In Japan, intercourse is "natural"; a kiss, pornographic. When Rodin's famous sculpture The Kiss came to Tokyo in the 1920s as part of a show of European art, it was concealed from public view behind a bamboo curtain.

Among cultures of the West, the number of nonsexual uses of the kiss is staggering: greeting and farewell, affection, religious or ceremonial symbolism, deference to a person of high status. (People also kiss icons, dice, and other objects, of course, in prayer, for luck, or as part of a ritual.) Kisses make the hurt go away, bless the sacred vestments, seal a bargain. In story and legend a kiss has started wars and ended them, awakened Sleeping Beauty and put Brunnhilde to sleep.

CLASSIFYING KISSES

Efforts to sort all of these kisses into neat categories apparently began centuries ago. According to Christopher Nyrop (1901), a Danish linguist who wrote a history of the kiss, the ancient rabbis recognized three kinds: greeting, farewell, and respect. The Romans also distinguished three kinds of kisses: *oscula* (friendly kisses), *basia* (love kisses), and *suavia* (passionate kisses). The most imaginative system was proposed in 1791 by an Austrian writer, W. von Kempelen, who divided kisses according to their sound: the *freundschaftticher hellklatschender Kerzenskuss* (the affectionate clear-ringing kiss coming from the heart), the acoustically weaker discreet kiss, and the *ekeljafter Schmatz* (a loathsome smack). Von Kempelen's categories did not gain widespread use.

When Nyrop wrote his book, he reported no fewer than thirty different German words to indicate types of kisses, from *Abschiedskuss* (farewell kiss) to *Zuckerkuss* (sweet, or "candy" kiss). The structure of the language permits composite nouns, but even so, German shows a remarkable linguistic richness in its variety of kisses. Today, *abkussen* means to give many little kisses all over; *erkussen* is (slang) for getting a gift or favor by kissing ("sucking up"?); *fortkussen* is to kiss away tears; and *wiederkussen* is to return a kiss you have been given.

The Germans are not the only ones to classify their kisses. Allan Edwardes, in *The Jewel in the Lotus* (1959), described the Hindu science of kissing: There is *sootaree-sumpoodeh* (the probing, tongue-sucking kiss), to be distinguished from *jeebh-juddh* (tongue-tilting), *jeebhee* (tongue scraping), and *hondh-chubbow* (lip biting). Of course, the question sexologists cannot yet answer—because research to date has been more descriptive than subjective—is to what extent Germans and Hindus actually have more diverse experiences than the French, who struggle along monolinguistically with *embrasser*, the Spanish, who have only besar, or the Russians, with *tselovat*.

CEREMONIAL KISSES

Classifications are entertaining but not especially illuminating. Types of kisses overlap and change over time. For example, St. Paul may not have known what would come of his simple advice to Christians to "salute one another with an holy kiss," a brief admonition in Romans 16:16 that is repeated in *Corinthians* I and II. Over the centuries, the "holy kiss" was interpreted and reinterpreted; it found expression in baptism, marriage, confession, and ordination. The *osculum pacis*, or kiss of peace, supposed to represent God's kiss of life and Christ's kiss of eternal blessing, was exchanged in some locations between priest and congregant, in others between clergy only. It passed out of common practice after the Reformation but is enjoying a renaissance in modern Catholic, Anglican, and Episcopalian congregations.

The famous kiss between bride and groom that concludes a wedding ceremony was actually part of ancient pagan rites and signified that legal bonds were being assumed. I've always thought the clergy's injunction to the groom "I now pronounce you husband [or 'man'] and wife, you may kiss the bride" represented the clergyman's quasi-parental permission to the new couple to be sexual (though it can't be an accident that the permission has typically been extended to the new husband!).

There are many more stories about holy and profane kisses, social and ceremonial kisses, changing customs (e.g., how European customs of kissing changed to bowing and hat lifting during the time of the plague), but by now I have made my point that an act like kissing cannot only be choreographed in very different ways but can serve many functions and carry many different meanings, all depending on the customs of a particular era. And, amazingly, each social group, each generation, feels its kisses are the normal, the natural ones.

IMPORTANCE OF KISSING

Why is kissing so popular, and why does it adapt to so many meanings? There have been some theories. Desmond Morris, following Freud, noted that a baby experiences its earliest joys, gratifications, and frustrations through its mouth, which becomes a site of emotional associations. In many cultures infants are lavishly touched, cuddled, and kissed all over their bodies, not only by their mothers but also by other relatives and friends. The infant learns that touching something soft with the mouth is a calming and pleasurable sensation. Adult kisses recall some of this infant gratification. Kissing symbols for luck, Morris argued, is emotionally reassuring—it's not just a random gesture to appease the gods or fate.

It certainly is true that the lips, mouth, and tongue are among the most exquisitely sensitive parts of the body. The tongue itself is sensitive to pressure, temperature, taste, smell, and movement. Lips, tongue, and mouth detect and transmit to the brain a range of incoming sensations; and the brain, in turn, devotes a disproportionate amount of its resources to processing their messages and linking them up to behavioral reactions and psychological functions. The space devoted to messages from and to the lips alone is far greater than that devoted to sensory or motor function for the entire torso.

Opportunities for kissing to develop multiple social meanings also arise because of variations in elements such as posture and facial expression—an especially important factor in communicating emotion. Through processes of social learning, trial and error, imitation, and reward and punishment, kisses acquire their multiple meanings and intense associations. Rules for social and sexual kisses vary not only across cultures but even among social classes and subcultures in the United States, making research into the social scripting of kissing a fertile area for those interested in how sexual choreography develops and what the various components mean to participants. The "naturalness" of kissing, as with so many other aspects of social life, turns out to be a biological potential shaped and cultivated by the real human nature—culture.

35 *The New Testament*: Letter of Paul to the Romans

PAUL, A SERVANT[a] OF JESUS CHRIST, called to be an apostle, set apart for the gospel of God [2]which he promised beforehand through his prophets in the holy scriptures, [3]the gospel concerning his Son, who was descended from David according to the flesh [4]and designated Son of God in power according to the Spirit of holiness by his resurrection from the dead, Jesus Christ our Lord, [5]through whom we have received grace and apostleship to bring about the obedience of faith for the sake of his name among all the nations, [6]including yourselves who are called to belong to Jesus Christ; [7]To all God's beloved in Rome,who are called to be saints: Grace to you and peace from God our Father and the Lord Jesus Christ.

[8]First, I thank my God through Jesus Christ for all of you, because your faith is proclaimed in all the world. [9]For God is my witness, whom I serve with my spirit in the gospel of his Son, that without ceasing I mention you always in my prayers, [10]asking that somehow by God's will I may now at last succeed in coming to you. [11]For I long to see you, that I may impart to you some spiritual gift to strengthen you, [12]that is, that we may be mutually encouraged by each other's faith, both yours and mine. [13]I want you to know, brethren, that I have often intended to come to you (but thus far have been prevented), in order that I may reap some harvest among you as well as among the rest of the Gentiles. [14]I am under obligation both to Greeks and to barbarians, both to the wise and to the foolish: [15]so I am eager to preach the gospel to you also who are in Rome.

[16]For I am not ashamed of the gospel: it is the power of God for salvation to everyone who has faith, to the Jew first and also to the Greek. [17]For in it the righteousness of God is revealed through faith for faith; as it is written, "He who through faith is righteous shall live."[b]

[18]For the wrath of God is revealed from heaven against all ungodliness and wickedness of men who by their wickedness suppress the truth. [19]For what can be known about God is

plain to them, because God has shown it to them. [20]Ever since the creation of the world his invisible nature, namely, his eternal power and deity, has been clearly perceived in the things that have been made. So they are without excuse; [21]for although they knew God they did not honor him as God or give thanks to him, but they became futile in their thinking and their senseless minds were darkened. [22]Claiming to be wise, they became fools, [23]and they exchanged the glory of the immortal God for images resembling mortal man or birds or animals or reptiles.

[24]Therefore God gave them up in the lusts of their hearts to impurity, to the dishonoring of their bodies among themselves, [25]because they exchanged the truth about God for a lie and worshipped and served the creature rather than the Creator, who is blessed forever! Amen.

[26]For this reason God gave them up to dishonorable passions. Their women exchanged natural relations for unnatural, [27]and the men likewise gave up natural relations with women and were consumed with passion for one another, men committing shameless acts with men and receiving in their own persons the due penalty for their error.

[28]And since they did not see fit to acknowledge God; God gave them up to a base mind and to improper conduct. [29]They were filled with all manner of wickedness, evil, covetousness, malice. Full of envy, murder, strife, deceit, malignity, they are gossips, [30]slanderers, haters of God, insolent, haughty, boastful, inventors of evil, disobedient to parents, [31]foolish, faithless, heartless, ruthless. [32]Though they know God's decree that those who do such things deserve to die, they not only do them but approve those who practice them.

[2]Therefore you have no excuse, 0 man, whoever you are, when you judge another; for in passing judgment upon him you condemn yourself, because you, the judge, are doing the same things. [2]We know that the judgement of God rightly falls upon those who do such things. [3]Do you suppose, 0 man, that when you judge those who do such things and yet do them yourself, you will escape the judgment of "God?[4]

End notes

a Or *slave*

b Or *The righteous shall live by faith*

1.1-7: Salutation. Ancient Greek letters customarily began with the names of the sender and of the recipient and a short greeting. Paul expands the usual form to express his Christian faith as well. **3-4:** God's *Son,* who came into the world physically *descended from David,* was manifested and installed in his true status at the resurrection. *The Spirit of holiness,* the Holy Spirit. **7:** *Saints,* those who belong to God, consecrated to his service. *Grace . . . and peace,* see 2 Th.1.2 n. .

1.8-15: Thanksgiving. After the salutation in ancient letters there usually came a short prayer of thanksgiving or of petition on behalf of the person addressed. This element also Paul expands in a characteristically Christian way.

1.16-17: The theme of the letter. In Christ God has acted powerfully to save men, offering righteousness and new life, to be received in faith. 17: The righteousness. of God is a state of pardon, or acceptance with God, which is not man's achievement but God's gift, originating in God's own righteous nature. *Through faith for faith*, faith is the sole condition of salvation. *He who ... shall live*, from Hab.2.4; compare Ga1.3.11; Phil.3.9; Heb.10.38.

1.18-32: God's judgment upon sin. 18: *Wrath*, see Co1.3.6 n. **19:** *What can be known*, i.e. apart from God's revelation to Israel and in Christ. **20-21:** Men have denied the knowledge of God that was given with their creation. **24,26,28:** *God gave them up*, because in turning from God they violated their true nature, becoming involved in terrible and destructive perversions; God has let the process of death work itself out. **29-31:** Ga1.5.19-21.

2.1-11: Jews are under judgement, as well as pagans (1.18-32).

36 *The Holy Qur'an*: Sura 27: "The Ants", Sura 29: "The Spider"

Sura 27. THE ANTS

A Meccan sura which takes its title from the ants mentioned in the Solomon story (verses 18-19). It both opens and closes by describing the Qur'an as joyful news for the believers and a warning for others. It gives stories of past prophets and the destruction of the communities that disbelieved in them. Illustrations are given of the nature of God's power, contrasted with the total lack of power of the 'partners' they worship beside Him, and descriptions are given of the Day of Judgement for those who deny it.

In the name of God, the Lord of Mercy, the Giver of Mercy

Ta Sin

These are the verses of the Qur'an—a scripture that makes things clear, a guide and joyful news for the believers who keep up the prayer, pay the prescribed alms, and believe firmly in the life to come. As for those who do not believe in the life to come, We have made their deeds seem alluring to them, so they wander blindly: it is they who will have the worst suffering, and will be the ones to lose most in the life to come. You [Prophet] receive the Qur'an from One who is all wise, all knowing.

Moses said to his family, 'I have seen a fire. I will bring you news from there, or a burning stick for you to warm yourselves.' When he reached the fire, a voice called: 'Blessed is the person near this fire[a] and those around it;[b] may God be exalted, the Lord of the Worlds. Moses, I am God, the Mighty, the Wise. Throw down your staff,' but when he saw it moving like a snake, he turned and fled. 'Moses, do not be afraid! The messengers need have no fear in My presence, I am truly most forgiving and merciful to those who do wrong[c] and then replace their evil with good. Put your hand inside your cloak and it will come out white, but unharmed. These are among the nine signs that you will show Pharaoh and his people; they have really gone too far.'

But when Our enlightening signs came to them, they said, 'This is clearly [just] sorcery!' They denied them—in their wickedness and their pride—even though their souls acknowledged them as true. See how those who spread corruption met their end!

We gave knowledge to David and Solomon, and they both said, 'Praise be to God, who has favoured us over many of His believing servants.' Solomon succeeded David. He said, 'People, we have been taught the speech of birds, and we have been given a share of everything: this is a clearly a great favour.' "Solomon's hosts of jinn, men, and birds were marshalled in ordered ranks before him, and when they came to the Valley of the Ants, one ant said, 'Ants! Go into your homes, in case Solomon and his hosts inadvertently crush you.' "Solomon smiled broadly at its words and said, 'Lord, inspire me to be thankful for the blessings You have granted me and my parents, and to do good deeds that please You; admit me by Your grace into the ranks of Your righteous servants.'

Solomon inspected the birds and said, 'Why do I not see the hoopoe? Is he absent? I will punish him severely, or kill him, unless he brings me a convincing excuse for his absence.' But the hoopoe did not stay away long: he came and said, 'I have learned something you did not know: I come to you from Sheba with firm news. I found a woman ruling over the people, who has been given a share of everything—she has a magnificent throne—[but] I found that she and her people worshipped the sun instead of God. Satan has made their deeds seem alluring to them, and diverted them from the right path: they cannot find the right path. Should they not worship God, who brings forth what is hidden in the heavens and earth and knows both what you people conceal and what you declare? He is God, there is no god but Him, the Lord of the mighty throne.' Solomon said, 'We shall see whether you are telling the truth or lying. Take this letter of mine and deliver it to them, then withdraw and see what answer they send back.'

The Queen of Sheba said, 'Counsellors, a gracious letter has been delivered to me. It is from Solomon, and it says, "In the name of God, the Lord of Mercy, the Giver of Mercy, do not put yourselves above me, and come to me in submission to God." ' She said, 'Counsellors, give me your counsel in the matter I now face: I only ever decide on matters in your presence.' They replied, 'We possess great force and power in war, but you are in command, so consider what orders to give us.' She said, 'Whenever kings go into a city, they ruin it and humiliate its leaders—that is what they do— but I am going to send them a gift, then see what answer my envoy brings back.'

When her envoy came to Solomon, Solomon said, 'What! Are you offering me wealth? What God has given me is better than what He has given you, though you rejoice in this gift of yours. Go back to your people: we shall certainly come upon them with irresistible forces, and drive them, disgraced and humbled, from their land.' Then he said, 'Counsellors, which of you can bring me her throne before they come to me in submission?' A powerful and crafty jinn replied, 'I will bring it to you before you can even rise from your place. I am strong and trustworthy enough,' but one of them who had some knowledge of the Book said, 'I will bring it to you in the twinkling of an eye.'

When Solomon saw it set before him, he said, 'This is a favour from my Lord, to test whether I am grateful or not: if anyone is grateful, it is for his own good, if anyone is ungrateful, then my Lord is self-sufficient and most generous.' Then he said, 'Disguise her throne, and we shall see whether or not she recognizes it.' When she arrived, she was asked, 'Is this your throne?' She replied, 'It looks like it.' [Solomon said], 'We were given knowledge before her, and we devoted ourselves to God—she was prevented by what she worshipped instead of God, for she came from a disbelieving people.' Then it was said to her, 'Enter the hall,' but when she saw it, she thought it was a deep pool of water, and bared her legs. Solomon explained, 'It is just a hall paved with glass,' and she said, 'My Lord, I have wronged myself: I devote myself, with Solomon, to God, the Lord of the Worlds.'

To the people of Thamud We sent their brother, Salih, saying, 'Worship God alone,' but they split into two rival factions. Salih said, 'My people, why do you rush to bring [forward] what is bad rather than good? Why do you not ask forgiveness of God, so that you may be given mercy?' They said, 'We see you and your followers as an evil omen.' He replied, 'God will decide on any omen you may see: you people are being put to the test.' There were nine men in the city who spread corruption in the land and did nothing that was good. They said, 'Swear by God: we shall attack this man and his household in the night, then say to his next of kin, "We did not witness the destruction of his household. We are telling the truth."' So they devised their evil plan, but We too made a plan of which they were unaware. See how their scheming ended: We destroyed them utterly, along with all their people. As a result of their evil deeds, their homes are desolate ruins—there truly is a sign in this for those who know—but We saved those who believed and were mindful of God.

We also sent Lot to his people. He said to them, 'How can you commit this outrage with your eyes wide open? How can you lust after men instead of women? What fools you are!' The only answer his people gave was to say, 'Expel Lot's followers from your town! These men mean to stay chaste!' We saved him and his family— except for his wife: We made her stay behind—and We brought [an awesome] rain down on them. How dreadful that rain was for those who had been warned!

Say [Prophet], 'Praise be to God and peace on the servants He has chosen.[d] Who is better: God, or those they set up as partners with Him? Who created the heavens and earth? Who sends down water from the sky for you—with which We cause gardens of delight to grow: you have no power to make the trees grow in them—is it another god beside God? No! But they are people who take others to be equal with God. Who is it that made the earth a stable place to live? Who made rivers flow through it? Who set immovable mountains on it and created a barrier between the fresh and salt water? Is it another god beside God? No! But most of them do not know. Who is it that answers the distressed when they call upon Him? Who removes their suffering? Who makes you successors in the earth? Is it another god beside God? Little notice you take! Who is it that guides you through the darkness on land and sea? Who sends the winds as heralds of good news before His mercy?[e] Is it another god beside God? God is far above the partners they put beside him! Who is it that creates life and reproduces it? Who is it that gives you provision from the heavens and earth? Is it another god beside God?' Say, 'Show me your evidence then, if what you say is true.'

Say, 'No one in the heavens or on earth knows the unseen except God.' They do not know when they will be raised from the dead: their knowledge cannot comprehend the Hereafter; they are in doubt about it; they are blind to it. So the disbelievers say, 'What! When we and our forefathers have become dust, shall we be brought back to life again? We have heard such promises before, and so did our forefathers. These are just ancient fables.' [Prophet], say, 'Travel through the earth and see how the evildoers ended up.' [Prophet], do not grieve over them; do not be distressed by their schemes. They also say, 'When will this promise be fulfilled if what you say is true?' Say, 'Maybe some of what you seek to hasten is near at hand.' Your Lord is bountiful to people, though most of them are ungrateful. He knows everything their hearts conceal and everything they reveal: there is nothing hidden in the heavens or on earth that is not recorded in a clear Book.

Truly, this Qur'an explains to the Children of Israel most of what they differ about, and it is guidance and grace for those who believe. Truly, your Lord will judge between them in His wisdom—He is the Almighty, the All Knowing—so [Prophet], put your trust in God, you are on the path of clear truth. You cannot make the dead hear, you cannot make the deaf listen to your call when they turn their backs and leave, you cannot guide the blind out of their error: you cannot make anyone hear you except those who believe in Our signs and submit to them. When the verdict is given against them, We shall bring a creature out of the earth, which will tell them that people had no faith in Our revelations. The Day will come when We gather from every community a crowd of those who disbelieved in Our signs and they will be led in separate groups until, when they come before Him, He will say, 'Did you deny My messages without even taking them in? Or what were you doing?' The verdict will be given against them because of their wrongdoing: they will not speak.

Did they not see that We gave them the night for rest, and the day for light? There truly are signs in this for those who believe. On the Day that the Trumpet sounds, everyone in heaven and on earth will be terrified—except such as God wills—and will come to Him in utter humility. You will see the mountains and think they are firmly fixed, but they will float away like clouds. This is the handiwork of God who has perfected all things. He is fully aware of…

Sura 29. THE SPIDER

A Meccan sura that takes its title from the illustration of the spider in verse 41. The sura stresses that believers will be tested and that they should remain steadfast. The misconceptions of disbelievers regarding the nature of revelation and the Prophet are addressed. References are made to earlier prophets and details given of the punishments brought on those who denied them.

In the name of God, the Lord of Mercy, the Giver of Mercy

Alif Lam Mim

Do people think they will be left alone after saying 'We believe' without being put to the test? We tested those who went before them: God will certainly mark out which ones are truthful and which are lying. Do the evildoers think they can escape us? How ill they judge!

But as for those who strive for their meeting with God, God's appointed time is bound to come; He is the All Seeing, the All Knowing. Those who exert themselves do so for their own benefit—God does not need His creatures—We shall certainly blot out the misdeeds of those who believe and do good deeds, and We shall reward them according to the best of their actions. We have commanded people to be good to their parents, but do not obey them if they strive to make you serve, beside Me, anything of which you have no knowledge: you will all return to Me, and I shall inform you of what you have done. We shall be sure to admit those who believe and do good deeds to the ranks of the righteous.

There are some people who say, 'We believe in God,' but, when they suffer for His cause, they think that human persecution is as severe as God's punishment—yet, if any help comes to you [Prophet] from your Lord, they will say, 'We have always been with you'. Does God not know best what is in everyone's hearts?—God will be sure to mark out which ones are the believers, and which the hypocrites. Those who disbelieve say to the believers, 'Follow our path and we shall take the consequences for your sins,' yet they will not do so—they are liars. They will bear their own burdens and others besides: they will be questioned about their false assertions on the Day of Resurrection.

We sent Noah out to his people. He lived among them for fifty years short of a thousand and they were doing evil when the Flood overwhelmed them. We saved him and those with him on the Ark. We made this a sign for all people.

We also sent Abraham. He said to his people, 'Serve God and be mindful of Him: that is better for you, if only you knew. What you worship instead of God are mere idols; what you invent is nothing but falsehood. Those you serve instead of God have no power to give you provisions, so seek provisions from God, serve Him, and give Him thanks: you will all be returned to Him. If you say this is a lie, [be warned that] other communities before you said the same. The messenger's only duty is to give clear warning.'

Do they not see that God brings life into being and reproduces it? Truly this is easy for God. Say, 'Travel throughout the earth and see how He brings life into being: He will bring the next life into being. God has power over all things. He punishes whoever He will and shows mercy to whoever He will. You will all be returned to Him. You cannot escape Him on earth or in the heavens; you will have no one to protect or help you besides God.' Those who deny God's Revelation and their meeting with Him have no hope of receiving My grace: they will have a grievous punishment.

The only answer Abraham's people gave was, 'Kill him or burn him!' but God saved him from the Fire: there truly are signs in this for people who believe. Abraham said to them, 'You have chosen idols instead of God but your love for them will only last for the present life: on the Day of Resurrection, you will disown and reject one another. Hell will be your home and no one will help you.' Lot believed him, and said, 'I will flee to my Lord: He is the Almighty, the All Wise.' We gave Isaac and Jacob to Abraham, and placed prophethood and Scripture among his offspring. We gave him his rewards in this world, and in the life to come he will be among the righteous.

And Lot: when He said to his people, 'You practise outrageous acts that no people before you have ever committed. How can you lust after men, waylay travellers, and commit evil in your gatherings?' the only answer his people gave was, 'Bring God's punishment down on us, if what you say is true.' So he prayed, 'My Lord, help me against these people who spread corruption.' When Our messengers brought the good news [of the birth of a son] to Abraham, they told him, 'We are about to destroy the people of that town. They are wrongdoers.' Abraham said, 'But Lot lives there.' They answered, 'We know who lives there better than you do. We shall save him and his household, except for his wife: she will be one of those who stay behind.' When Our messengers came to Lot, he was troubled and distressed on their account. They said, 'Have no fear or grief: we shall certainly save you and your household, except for your wife—she will be one of those who stay behind—and we shall send a punishment from heaven down on the people of this town because they violate [God's order].' We left some [of the town] there as a clear sign for those who use their reason.

To the people of Midian We sent their brother Shu'ayb. He said, 'My people, serve God and think ahead to the Last Day. Do not commit evil and spread corruption in the land.' "They rejected him and so the earthquake overtook them. When morning came, they were lying dead in their homes. [Remember] the tribes of 'Ad and Thamud: their history is made clear to you by [what is left of] their dwelling places. Satan made their foul deeds seem alluring to them and barred them from the right way, though they were capable of seeing. [Remember] Qarun and Pharaoh and Haman: Moses brought them clear signs, but they behaved arrogantly on earth. They could not escape Us and We punished each one of them for their sins: some We struck with a violent storm; some were overcome by a sudden blast; some We made the earth swallow; and some We drowned. It was not God who wronged them; they wronged themselves. Those who take protectors other than God can be compared to spiders building themselves houses—the spider's is the frailest of all houses—if only they could understand. God knows what things they call upon beside Him: He is the Mighty, the Wise. Such are the comparisons We draw for people, though only the wise can grasp them. God has created the heavens and earth for a true purpose. There truly is a sign in this for those who believe.

[Prophet], recite what has been revealed to you of the Scripture; keep up the prayer: prayer restrains outrageous and unacceptable behaviour. Remembering God is greater: God knows everything you are doing. [Believers], argue only in the best way with the People of the Book, except with those of them who act unjustly. Say, 'We believe in what was revealed to us and in what was revealed to you; our God and your God are one [and the same]; we are devoted to Him.' This is the way [Muhammad] We sent the Scripture to you. Those to whom We had already given Scripture[g] believe in [the Qur'an] and so do some of these people. No one refuses to acknowledge Our revelations but those who reject [the Truth].

You never recited any Scripture before We revealed this one to you; you never wrote one down with your hand. If you had done so, those who follow falsehood might have had cause to doubt. But no, [this Qur'an] is a revelation that is clear to the hearts of those endowed with knowledge. Only the evildoers refuse to acknowledge Our revelations. They say, 'Why have no miracles been sent to him by his Lord?' Say, 'Miracles lie in God's hands; I am simply here to warn you plainly.' Do they not think it is enough that We have sent down to you the Scripture that is recited to them? There is a mercy in this and a lesson for believing people. Say, 'God is sufficient witness between me and you: He knows what is in the heavens and earth. Those who believe in false deities and deny God will be the losers.'

They challenge you to hasten the punishment: they would already have received a punishment if God had not set a time for it, and indeed it will come to them suddenly and catch them unawares. They challenge you to hasten the punishment: Hell will encompass all those who deny the truth, on the Day when punishment overwhelms them from above and from below their very feet, and they will be told, 'Now taste the punishment for what you used to do.'

My believing servants! My earth is vast, so worship Me and Me alone. Every soul will taste death, then it is to Us that you will be returned. We shall lodge those who believed and did good deeds in lofty dwellings, in the Garden graced with flowing streams, there to remain. How excellent is the reward of those who labour, those who are steadfast, those who put their trust in their Lord! How many are the creatures who do not have their sustenance stored up: God sustains them and you: He alone is the All Hearing, the All Knowing. If you ask the disbelievers who created the heavens and earth and who harnessed the sun and moon, they are sure to say, 'God.' Then why do they turn away from Him? It is God who gives abundantly to whichever of His servants He will, and sparingly to whichever He will. He has full knowledge of everything. If you ask them, 'Who sends water down from the sky and gives life with it to the earth after it has died?' they are sure to say, 'God.' Say, 'Praise be to God!' Truly, most of them do not use their reason.

The life of this world is merely an amusement and a diversion; the true life is in the Hereafter, if only they knew. Whenever they go on board a ship they call on God, and dedicate their faith to Him alone, but once He has delivered them safely back to land, see how they ascribe partners to Him. Let them show their ingratitude for what We have given them; let them take their enjoyment. Soon they will know. Can they not see that We have granted them a secure sanctuary[h] though all around them people are snatched away? Then how can they believe in what is false and deny God's blessing? Who could be more wicked than the person who invents lies about God, or denies the truth when it comes to him? Is Hell not the home for the disbelievers? "But We shall be sure to guide to Our ways those who strive hard for Our cause: God is with those who do good.

End notes

a Moses or God. Literally 'in this fire' (*fi al-nari*). Zamakhshari interprets *fi* as 'near', while Qatada and Zajjaj understand *nar* 'fire' to mean *nur* 'light' (Razi).

b The angels.

c Cf. 28: 15. This is an allusion to a man Moses killed in Egypt.

d As messengers, see e.g. 3: 33.

e i.e. rain.

f Cf. e.g. 51:28.

g According to some commentators this refers to those Jews and Christians at the time of the Prophet who believed in him.

h Mecca.

37 *The Morality of Homosexuality*

Michael Ruse

IS HOMOSEXUALITY BAD SEXUALITY BECAUSE IT IS BIOLOGICALLY UNNATURAL?

A key philosophical charge against homosexual activity is that it is "unnatural." "Nature," in this context, is intended to refer to our *biological* nature. Hence, the conclusion is drawn that such activity should be condemned as immoral. This was the cry of Plato, it was echoed by Aquinas and Kant, and it is still with us today. There are a number of points I want to make about this argument.

First, if we mean by "homosexuality is unnatural" that it is never a practice to be found in the animal world—and this is certainly what Plato thought—then the claim is simply false. There is a substantial body of evidence that supports the conclusion that homosexual activity is widespread throughout the animal world. Virtually every animal whose activity has been studied in detail shows some forms of homosexual behavior. Mutual masturbation, anal intercourse, and so forth, are commonplace in the primate world. Similarly, amongst other mammals, we find all sorts of activity that can only truly be spoken of as "homosexual," in some sense. One male will mount another and come to climax. Analogously, females show deep bonds and sexual type behavior towards each other. Sometimes this behavior of animals is manifested just in young animals. In other cases, the homosexual activity is ongoing, if not exclusive (see Weinrich 1982 for details and references).

If homosexual activity is so widespread in the animal world, why has it not been noted before? In fact, it has been noted before; but, with their usual selectivity, people writing on human sexuality have failed to note it or have simply been ignorant about it. Then again, there has been such a fixed notion that homosexuality belongs exclusively to the human world, that people simply have not been able to see animal homosexuality— even when they have been presented with the clearest evidence of it.

A revealing example of this selective vision is given by a recent researcher on mountain sheep. He wrote two books on the subject: one in 1971 and the other a short time later

(Geist 1971; 1975). In the first book, there was a great deal of discussion of male dominance and pecking order, with alpha males fighting and subduing beta males. In the second book, the author came right out and said what he had been seeing all along was homosexual activity. The alpha males mount the beta males, having erections and emissions, sometimes involving anal intercourse. Candidly, the author admitted that he simply had not been able to bring himself to think of this as homosexuality. In his own words: "Those magnificent animals, queers?" This case is atypical only in that the researcher was more candid than most. If one's desire is to argue that homosexuality is unnatural, the animal world is certainly not the place to look.

A second pertinent point is that homosexuality might have a biological function in humans. The basic mechanism of the central biological theory of Darwinian evolution is "natural selection"; or "the survival of the fittest." Those organisms more successful at reproducing than others pass on their units of inheritance—their genes—and are thus the organisms most represented in the next and future generations. Given enough time, this process leads to full-blown evolution.

Reproduction is the key to evolutionary success. It is possible, however, to reproduce by proxy. Suppose that, instead of reproducing one's self, one aids close relatives to reproduce more efficiently. Then, in a sense, one is increasing the representation of one's own units of inheritance in the next and future generations, simply because one shares these units of inheritance with close relatives. This vicarious reproduction is known as "kin selection," and it has been very extensively documented in the animal world, particularly in the hymenoptera (ants, bees, and wasps). (For a quick introduction to Darwinism, see Ruse 1982. For more on kin selection, see Wilson 1975 or Dawkins 1976.)

Kin selection in humans provides a possible biological explanation of the homosexual lifestyle, as an alternative reproductive strategy. Recent research has shown that, in non-industrial societies, male homosexuals frequently fit a pattern one would expect were kin selection operating (Weinrich 1976). These males would probably not be efficient as direct reproducers, because of such factors as debilitating childhood illnesses. They do, however, hold positions in society that can significantly aid close relatives. For example, in many American Indian tribes homosexuals take on the role of the shaman—that is, of a kind of magical figure who has to be consulted by the tribe before great events like battles can take place. The shaman has considerable power and financial influence within the community. He is, thus, in a strong position to aid close relatives (siblings, nephews, nieces, and so on). There is, moreover, evidence that such help actually occurs. It seems plausible to suppose, therefore, that in such cases biology itself has promoted genes for manifesting homosexual inclinations and activity.

If indeed kin selection, or some like process, does operate in a way such as that just suggested, then biology is at least a partial cause of human homosexuality. It would therefore be odd to speak of homosexuality as being "unnatural." If by "natural," you mean that which nature has done, homosexuality would be as natural as heterosexuality. Indeed, forcing homosexuals to live heterosexual lifestyles would be unnatural from a biological point of view, not the converse. (The evidence for the biological foundations of human homosexuality is presented in detail, together with various explanatory models, in Ruse 1981.)

A third point is so obvious that it is usually overlooked. We humans do not live in a world of strict biology. We are cultural creatures, which is why we are so successful as a species. We have speech, customs, religion, literature, art—and even philosophy. To speak of humans as "just animals" is to ignore half the story. Any evaluation of human homosexuality from a natural or unnatural perspective must, therefore, take our culture into account. The fact that we do not always do the things that animals do, does not mean that it is unnatural for us not to do such things. It is simply a reflection of the fact that, by nature, we are not as other animals (Lumsden and Wilson 1981).

Hence, if homosexual activity is part of human culture—and it certainly is in many respects— then to speak of it as "unnatural," judged purely from a physiological perspective, is simply meaningless. It could indeed be true that animals do not practice homosexual activity (although, it so happens that it is not true); it could be true that humans do, in fact, practice homosexual activity (as indeed they do); but, the conclusion would not necessarily be that human homosexuality is unnatural. The conclusion could simply be that such behavior is part of our human nature and not part of animal nature.

Finally, let me point out that even if homosexuality were biologically unnatural, this need not make it immoral. Because we do something which is against our biological nature, it does not follow that the act is wrong. As many philosophers have pointed out, to argue from what is (i.e., biological nature) to what *ought* to be (i.e., the morally desirable) is a fallacy. Whether we do something or not is one thing. Whether it is moral or immoral is quite another. The two are not logically connected.

There are, then, four independent rejoinders to the "unnatural" argument that has dominated both classical and modern discussions of homosexuality: it is false that animals are not homosexual; it is false that homosexuality must be antireproductive and nonbiological; it is false that homosexuality is to be judged without taking note of the cultural nature of humans; and it is false that what is unnatural is necessarily immoral.

THE KANTIAN ANALYSIS

Kant was strongly against homosexuality. Obviously, apart from any religious biases Kant may have had, whatever he himself may have thought, essentially this opposition was based on the homosexuality-as-unnatural thesis. Assuming that the arguments of the last sections are effective, what then of the basic Kantian philosophy? Can one indulge in homosexual activity and yet be true to the Categorical Imperative? Kant himself did not think so. Nevertheless, my own sense of sex in general, and of homosexuality in particular, is that once St. Paul and Plato are put aside, the Categorical Imperative is far less of an impediment to variant sex than its author supposed (see also Baumrin 1975).

The Categorical Imperative demands that people be treated as ends, and not as means only. "Act so that you treat humanity, whether in your own person or in that of another, always as an end and never as a means only" (Kant 1959, p. 47). Nothing in the Imperative itself rules out the possibility of the relationship being a homosexual one, rather than heterosexual. Homosexuals, male or female, fall in love with partners and, under any meaningful sense of the term, treat those partners as ends and not simply as means. Thus, homosexual activity as such is not ruled out by the Categorical Imperative. Indeed, in the right circumstances, a Kantian should rather think that one ought to behave homosexually. (Suppose, for instance, one were faced with a choice of would-be partners, one of the same sex and one of the other sex, and one was oneself drawn homosexually to the same sex partner—and to act otherwise would involve deceit and unkindness.)[1]

THE UTILITARIAN ANALYSIS

Turning to the other great moral theory, let us first distinguish between the two main versions of utilitarianism. There is that associated with the name of Jeremy Bentham (1948), and there is that associated with the name of John Stuart Mill (1910). Benthan argued that the utility against which the consequences of all acts should be judged is any kind of pleasure that one finds desirable. He made no distinction between various pleasures or happinesses. John Stuart Mill, on the other hand, argued strongly that one can grade pleasures and happinesses and that some are much to be preferred to others. In particular, Mill argued that the more intellectual sorts of pleasures are more worthwhile. "Better to be Socrates dissatisfied than a fool satisfied" (Mill 1910, p. 9).

A Benthamite utilitarian appears to endorse homosexual activity just as did Bentham himself. If an individual enjoys homosexual activity, then it is a good thing for that person and she/he should strive to maximize its occurrence. Moreover, assuming that others enjoy it also, he/she should strive to let them enjoy it to the full.

What about Mill? I am not sure that a Millian would be quite as easy about sex as a Benthamite; but with regard to homosexual activity as such, the conclusion seems similar. Certainly, a Mill-type utilitarian would think that activity within a loving relationship, whether heterosexual or homosexual, was a good thing and ought to be promoted. Indeed, as with the Kantian analysis, there will be cases where a Millian (as well as a Benthamite) could urge homosexual activity on someone—not to show affection and not to act homosexually would be wrong.

IS HOMOSEXUALITY PERVERTED?

Is homosexual behavior perverted behavior? With Nagel's critics, I would argue that he described something better called "incomplete" sex than "perverted" sex. But, like Nagel, I doubt that the notion of completeness in itself throws much moral light on sexuality—certainly, it throws no more light than that gained from the traditional moral theories just discussed. If some sex is bad sex, it is not so much because it is incomplete; rather, it is because the sex violates the Categorical Imperative or fails to lead to true happiness.

Does this mean that the notion of sexual "perversion" is an empty one? Not at all. Despite my earlier strictures about the concept of "naturalness" as it occurs in the philosophical literature, I do not want to deny that in some sense the perverted is the nonnatural or the unnatural. But, for reasons given, naturalness cannot be defined in terms of pure biology, as argued by philosophers from Plato to Ruddick. Humans are cultural beings, and what is natural must be understood in terms of culture. Hence, the unnatural—the perverted—is something that goes against cultural norms.

What exactly does this mean? A purely statistical definition, as offered by Goldman, will not do. Being in a minority is not as such nonnatural or perverse. Rather, nonnatural sex is sex that goes against our personal nature as cultural beings. It is something that we simply would not want to do even if we could. Remember the story of Gyges in Plato's *Republic*. He was the fellow who found a ring that enabled him to become invisible. As a result of this, he seized power in the kingdom, killing the king and seducing the queen. We can all understand what Gyges was up to, and even though we may not approve of his actions, we do not find them absurd or weird or nonnatural.

Unnaturalness is something that we simply would not want to do, even if we had gotten Gyges's ring. More particularly, it is something we could not imagine wanting to do. I might not want to murder, even though the ring enables me to do it. I can imagine playing Gyges's role, however. Suppose that, thanks to the ring, I could now spend my days concealed in the corner of a public lavatory, watching and smelling people defecate. I cannot imagine putting the ring to that use. In short, that activity for me, would be unnatural. It would be a "perversion." (Note that not all perversions are necessarily sexual. I am not sure that watching folks defecate would be.)

Thus, I argue that the perverse is the unnatural, where the unnatural is that which goes against what an individual finds culturally comprehensible.[2] The perverse is that which one cannot even conceive of wanting to do, even if one could. Now, this raises a number of questions. First, does the notion of perversion, as defined, have any value connotations? Second, would such value connotations (if they exist) necessarily be moral value connotations? Third, where does this leave the question of homosexual activity? Let me take these questions in turn, briefly.

First, the notion of perversion does have strong value connotations. There is little doubt (except perhaps in minds of academic philosophers), that when we speak of something as "perverted," we mean that it is in some way vile or disgusting. Just above, I gave an example of something I consider a perversion. Would it surprise you to learn that I found it difficult simply to write it down? It shouldn't! It's a perversion! In my opinion, hanging around the lavatories, watching and smelling others defecate, is a thoroughly disgusting thing to do. That is the whole point of my reference to Gyges and his ring. Watching and smelling people defecate is something that I cannot even conceive of wanting to do. I do not want to steal a camera from my colleague's office, but I can certainly conceive of situations where I might do something like this. I do not find the thought of stealing a friend's camera disgusting, although I do find it shameful. Hence, there are strong value connotations involved in the notion of perversion or non-naturalness.

Second, what of the connection between perversion and morality? Undoubtedly, that which is perverted is often immoral. For instance, strangling a small child and simultaneously raping her is both perverted and grossly immoral. However, this does not mean that the notions of perversion and morality are logically connected. Certainly, that which is immoral is not always perverted. Stealing a colleague's camera would be an immoral act; but it would not be a perverted act.

Are all perverted acts immoral acts? I doubt it. Would my example above be immoral? Perhaps you would think it an invasion of privacy. But what if I waited until the lavatories were empty, and then went in to drink from the urinals? This would be perverted; but, I'm not at all sure that it would be immoral. Of course, the simple fact of the matter is that that which is perverse is often immoral because many things that we cannot even conceive of wanting to do would be immoral things to do. Any connections of this type are contingent rather than logical. Hence, although perversion has with it an element of disgust, which does surely involve values, perversion in itself does not entail a moral repulsion. Therefore, although those who argue that perversion is in some sense bad are right, they are also right when they argue that this badness is not in itself a moral badness. It is more something akin to an aesthetic badness.[3] We are revolted by perversions, but this does not necessarily entail moral condemnation.[4]

Third and finally, what about the connection between homosexuality and perversion? Note that the way in which I have characterized "perversion" makes it a subjective phenomenon, which could vary from person to person within a culture. Some people find certain things revolting, others do not. Expectedly, the same goes for perversions. Thus, for instance, I do not find oral sex a particularly revolting phenomenon. On the other hand, other people find it thoroughly disgusting. For me, therefore, oral sex is not a perversion. For others, oral sex clearly is a perversion.

What does all this mean for homosexuality, especially as it applies to contemporary North America? The answer is relativistic. For some people, homosexuality is indeed a perversion. They find it disgusting and recoil from the very thought of it. Others (not necessarily just homosexuals) do not find homosexual activity a perversion. This is not to say that everybody who finds homosexuality not to be perverted, wants to behave homosexually. But, it is to say that such people could, in some sense, imagine freely doing it—at least, they can put themselves in the place of someone who would want to do it. They are certainly not overwhelmed by a sense of disgust.[5]

Hence, I argue that there is no straightforward answer to the question of whether or not homo-sexuality is a perversion. Some people regard it as such; others do not. Clearly, we have had something of a change in the last fifty years, with fewer people now thinking homosexual activity perverted. Perhaps, we will continue to see a change. As things stand at the moment, for some people homosexuality is a perversion, and this is all there is to be said on the matter. As a consequence of this, for some people in our society, homosexuality is not the best kind of sex. For them, it is aesthetically inferior (more bluntly, it is revolting sex). But this is not to say that those who think this way are therefore justified in inferring that homosexuality is immoral. (Although, I am sure that many do, in fact, conclude this.) For others, homosexuality is not a perversion; it is not in any sense inferior or worse sex.

IS HOMOSEXUALITY BAD SEXUALITY?

Let us list the conclusions. First, it is simply bad science to go on arguing that human homosexual activity is biologically unnatural. Even if it were, this would tell us nothing of its moral status.

Second, although religion is *prima facie* hostile to homosexuality (and probably truly hostile), there is nothing in the standard philosophical theories of moral behavior that outrightly condemns homosexual activity. Both Kantians and utilitarians can and should approve of homosexual activity *per se*.

Third, the notion of perversion properly understood does have value connotations; namely, negative values or revulsion and disgust. Undoubtedly, many people in our society do find homosexual activity revolting. For them it is a perversion. However, this is not a universal feeling. There are heterosexuals, as well as homosexuals, who do not look upon such activity as a perversion. For them, homosexuality is perfectly good sexuality.

In response to the title of this essay, therefore, I reply that, in important respects, homosexuality is certainly not bad sexuality. It is perfectly good sexuality. However, many people look upon homosexual activity as a perversion. That is a fact, and no amount of empathetic philosophizing can change this. This is not to say, however, that the possibility of changing people's opinions is not open to us all. As philosophers, caring about human beings, aware of how much hatred today is directed towards homosexuals, we have a special obligation to work toward such change.

End notes

1. In this discussion, I am simply considering basic homosexual activity, essentially between two people who have some feeling for each other. I am not considering possible moral complications, like group sex, since presumably they are not distinctive homosexual matters. However, I must note that male homosexuals are often given to highly promiscuous behavior (Bell and Weinberg, 1978). Were one to think this a moral issue, then as a matter of fact this would be of particular concern in a full analysis of homosexual activity as it typically occurs. See Elliston (1975) for a spirited argument for the value of promiscuity; although, he does qualify his enthusiasm by allowing the promiscuity as a "limited value in the movement toward a sexual idea," which ideal involves "a full commitment to a single other" (p. 240), I am not sure that much male, promiscuous homosexual activity is properly viewed as moving towards such an ideal. Also, current health threats, such as Acquired Immune Deficiency Syndrome (AIDS), ought to be considered. See Silverstein and White 1977, and *Science 83* (April 1983), for more on these matters.
2. In this essay, I will just deal with perversion at the individual level, where society as a whole is undecided on the subject. This seems to me to be the case for homosexuality in our society today. But, a full analysis of perversion would need reference to when a society as a whole could be said to consider a practice "perverted." This would clearly need discussion of the way most people felt, but would probably also need to refer to other societal beliefs, like religion.
3. A philosophical emotivist could argue that, since ethics is all feelings, morality and perversion collapse together. Since I am not a philosophical emotivist in any usual sense, this is not my worry.
4. If, as Kant argues, one has obligations to oneself, then it might be argued that one ought not degrade oneself by performing that which one finds perverted. But, as will be seen, it does not follow that one must judge others immoral for doing what they do and not find it perverted, even though the individual judging does find it to be immoral. Nor, speaking now as an enthusiast for Mill's views on liberty, does it follow that one should at once try to stop that which one would judge perverted. I thus disagree with Devlin's (1965) views about the propriety of legislating on private practices; although, as a matter of fact Devlin himself spoke out against antihomosexual laws.

5. In associating perversion with unnaturalness and disgust, my thinking parallels Slote (1975). However, I see no reason to argue as he does that "the kinds of acts people call unnatural are those that most people have some impulse toward that they cannot or will not admit to having" (p. 263). This may often be true, and perhaps in the case of homosexuality explains the violent emotions engulfing many homophobics. But I deny having even subconscious coprophilic tendencies.

Acknowledgements

"The Ants, The Spider" in *The Holy Qur'an*, translated by M.A.S. Abdel Haleem, pp. 239-243, 252-256. © 2004 Oxford University Press.

Aquinas, Thomas. Excerpt from *Summa Theologica*, edited by Edward Batchelor, pp. 39-47. © 1980 The Pilgrim Press.

Aristotle. Excerpt from *Nicomachean Ethics*, translated by Terrence Irwin, pp. 207-224. © 1985 Hackett.

Blackburn, Simon. "Chapter 10: Hobbesian Unity and Chapter 11: Disasters" in *Lust: The Seven Deadly Sins*, pp. 87-102. © 2004 Oxford University Press.

Burtt, E.A. "The Bodhisattva's Vow of Universal Redemption" in *The Teachings of the Compassionate Buddha*, pp. 130-135. © 1982 Mentor.

Catullus. "Poems 5, 7, 32, 75" in *Lyrics Rude & Erotic*, translated by Ewan Whyte, pp. 11, 15, 37, 121. © 2004 Mosaic Press.

Confucius. Book 1 of *The Analects*, pp. 59-62. © 1979 Penguin.

Excerpt from *The Hebrew Bible*, pp. 1-5, 19-21. © Oxford University Press.

Excerpt from *The Hebrew Bible*, pp. 815-21. © Oxford University Press.

Excerpt from *The New Testament*, pp. 1380-1381, 1392-1393. © Oxford University Press.

Excerpt from *The New Testament*, pp. 145-146. © Oxford University Press.

Frankfurt, Harry G. Excerpt from *The Reasons to Love*, pp. 35-51. © 2004 Princeton.

Freud, Sigmund. Excerpts from *Civilization and its Discontents*, pp. 48-49, 55-63. © 1961 W.W. Norton.

Fromm, Eric. 1956. "Erotic Love" in *Philosophy of Sex and Love*, edited by Robert Trevas, Arthur Zucker, and Donald Borchert, pp. 53-54. © 1997 Prentice Hall.

Haslett, Adam. "Love Supreme" in *The New Yorker*, pp. 76-80. © May 31, 2004 CondeNet.

Hesiod. Excerpts from *Theogony*, translated by Stanley Lombardo, © 1914 Hackett Publishing Company, Inc..

Nagel, Thomas. "Sexual Perversion" in *Mortal Questions*, pp. 37-52. © 1979 Cambridge University Press.

Nietzsche, Friedrich. Excerpt from *The Gay Science*, edited by Ellen Feder, Karmen MacKendrick and Sybol Cook, pp. 477-478. © 2004 Pearson/Prenctice Hall.

Olds, Sharon. "Sex Without Love" in www.poemhunter.com/p/m/poem.asp?poet=10652&poem=106308

Plato. "Symposium: Speech of Aristophanes" in *Symposium*, translated by Alexander Nehamas and Paul Woodruff, pp. 25-31. © 1989 Hackett.

Plato. "Symposium: Speech of Socrates" in *Symposium*, translated by Alexander Nehamas and Paul Woodruff, pp. 50-60. © 1989 Hackett.

Plato. Excerpt from *Republic*: Book VII, translated by G.M.A Grube, pp. 167-171. © 1992 Hackett.

Ruse, Michael., 1984. "The Morality of Homosexuality" in *Philosophy and Sex and Love*, edited by Robert Trevas, Arthur Zucker and Donald Borchert, pp. 262-268. © 1997 Prentice Hall.

de Sade, Marquis. Excerpt from Philosophy in the Bedroom in *A Passion for Wisdom*, edited by Ellen Feder, Karmen MacKendrick, and Sybol Cook, pp. 397-403. © 2004

Sappho. "Fragments" in *A Passion for Wisdom*, edited by Ellen Feder, Karmen MacKendrick and Sybol Cook; Anne Carson (trans.), pp. 8-10. © 2004 Pearson/Prentice Hall.

Shakespeare, William "Sonnet XXIX" in *Complete Sonnets*, pp. 13-14. © 1991 Dover.

Sinclaire Intimacy Institute "Human Sexual Response Cycle" in http://www.health.discovery.com/centers/sex/sexpedia/sexresponse_print.html, pp. 1-3. © May 18, 2005 Sinclaire Institute.

Soble, Alan. "Varieties of Love" in *The Philosophy of Sex and Love*, © 2008 Paragon House Publishers.

St. Augustine. Excerpts from *Confessions*, pp. 43-53, 55-62, 211-213, 231-250. © 1982 Penguin.

St. Teresa of Avila. "The Life of Saint Teresa by Herself" in *A Passion for Wisdom*, edited by Ellen Feder, Karmen MacKendrick, and Sybol Cook, pp. 191-199. © 2004 Prentice Hall.

Tiefer, Leonore. "Historical, Scientific, Clinical, and Feminist Criticisms of 'The Human Sexual Response Cycle' Model" in *Sex is Not a Natural Act*, pp. 41-58. © 1995 Westview Press.

Tiefer, Leonore. "The Kiss" in *Sex is Not a Natural Act*, pp. 77-81. © 1995 Westview Press.

"Women" in *The Holy Qur'an*, translated by M.A.S. Abdel Haleem, © 2004 Oxford University Press.

Yeats, W.B. "For Anne Gregory".